T0073902

The Musical Brain

The Musical Brain

What Students, Teachers, and Performers Need to Know

LOIS SVARD

OXFORD
UNIVERSITY PRESS

Oxford University Press is a department of the University of Oxford. It furthers
the University's objective of excellence in research, scholarship, and education
by publishing worldwide. Oxford is a registered trade mark of Oxford University
Press in the UK and certain other countries.

Published in the United States of America by Oxford University Press
198 Madison Avenue, New York, NY 10016, United States of America.

© Oxford University Press 2023

Library of Congress Cataloging-in-Publication Data
Names: Svard, Lois, author.
Title: The musical brain : what students, teachers, and performers need to know / Lois Svard.
Description: New York : Oxford University Press, 2023. | Includes index.
Identifiers: LCCN 2022040680 (print) | LCCN 2022040681 (ebook) |
ISBN 9780197584170 (hardback) | ISBN 9780197584194 (epub)
Subjects: LCSH: Music—Psychological aspects. | Music—Physiological aspects. |
Musical ability. | Music—Origin.
Classification: LCC ML3830 .S948 2023 (print) | LCC ML3830 (ebook) |
DDC 781.1/1—dc23/eng/20220826
LC record available at https://lccn.loc.gov/2022040680
LC ebook record available at https://lccn.loc.gov/2022040681

DOI: 10.1093/oso/9780197584170.001.0001

1 3 5 7 9 8 6 4 2

Printed by Sheridan Books, Inc., United States of America

Contents

Preface

We make and listen to music for the powerful effect it has on our emotions, and we can't imagine our lives without it. Yet we tend to know nothing about the intricate networks that billions of neurons create throughout our brains to make music possible. Neuroscientists, however, have been studying musicians and the process of making music for the past thirty years and have accumulated an overwhelming amount of information about brain processing for music. Although many of these neuroscientists are passionately interested in music and some are musicians themselves, their interests and expertise are in the functioning of the brain. As one neuroscientist told me, it will need to be a musician who applies this research to practice and performance. Hence this book—written by a musician for people who love music, who make music, and who want to know more about the brain processes that not only make it possible but can help all of us to become better students, teachers, performers, or listeners.

This book is an exploration of the research relevant to how and when music study should begin; how to practice, learn, and perform; and why experiencing music together is so fundamental to human lives. It is written from the viewpoint of a pianist who has performed a great deal, taught hundreds of students with varying degrees of technical expertise and musicality, and delved deeply and with fascination into the interrelationships between neuroscience and music. Every topic covered in the book could be, and often has been, the subject of tens to hundreds of research articles and sometimes a full-length book or two. I have written earlier about some of these topics on my blog, *The Musician's Brain*, but these subjects are dealt with here in more detail and discussions include more recent research.

The first three chapters lay the groundwork for considering music as a fundamental part of who we are as human beings—biological foundations for music. They explore research that shows separate brain areas for language and music, look at evidence for music dating back tens of thousands of years, and consider the many musical abilities we have at birth.

The next four chapters explore the marvelous improvisational ability of our brains to change in response to experience, learning, or injury. This is called neuroplasticity, and it makes it possible to learn an instrument and learn music. An injured brain can even create new pathways. Cross-modal neuroplasticity, the extraordinary ability of the brain to repurpose sensory processing areas, makes it possible for blind or deaf people to become highly skilled musicians. Unfortunately, neuroplasticity can also be maladaptive, causing a movement disorder called focal dystonia that makes it impossible for a musician to play his instrument. Although neuroplasticity is the cause of this disorder, it can sometimes also be a vehicle for recovery because of the brain's ability to adapt.

Two further chapters explore other remarkable brain processes that support music making: imagery—especially motor imagery and its significant implications for learning music, and mirror neurons—those astonishing brain neurons that fire both when we perform an action and when we see that same action being performed, thus having a major impact on learning, teaching, and performance. And the last chapter considers the thirty-year-old question, "Does music make you smarter?" While some obvious claims can be discounted, there exists a great deal of evidence to demonstrate that learning to play a musical instrument does, in fact, have cognitive benefits that last a lifetime.

Research has shown us that music was an important part of the lives of our prehistoric ancestors tens of thousands of years ago. The fact that our brains have evolved to support music-making, from the most basic level to highly skilled performance, tells us how important music is to human life. Making music isn't just something we do; it is part of who we are as human beings.

Acknowledgments

I would like to extend my deepest gratitude to the following:

Sandra Blakeslee for introducing me to the art of science writing; Kyle Gann for his inspiring example of writing about music and musicians—and also for preparing the musical examples for the book.

The countless number of researchers who have made the exploration of neuroscience and music the focus of their work. Their fascinating research makes this book possible.

Neuroscientists Nina Kraus and Eckart Altenmüller, and cognitive psychologist Laurel Trainor for speaking with me at length about their research, for reading early drafts concerning their work and providing clarifications, and for allowing use of photos or images in the book and on the website.

Musicians Jake Heggie, Ann Schein, Nina Skolnik, Lynn Eustis, Jonathan Palant, David Vining, and Jim Litzelman for being willing to share personal experiences that helped shape the narrative of the book.

Margaret Martin, for her infectious enthusiasm about Harmony Project and how music can change the world.

Fei-hsing Hsu and Hsing-ay Hsu for sharing memories of their brother and uncle, Fei-Ping Hsu; Anne Pusey for her translations of Mandarin Chinese text into English; Than Mitchell and Ann Keeler Evans for added insights into Steve Mitchell's life.

Holly Fischer for her marvelous images of the brain; Jim Litzelman and Lea Pearson for allowing the use of their videos and print materials on the website; the many individuals and organizations who provided wonderful photographs. Their names are credited in the photo captions.

The wonderful friends who read all or portions of the manuscript: Lynn Eustis, for her invaluable insight on all matters having to do with the voice; Kay Hooper, for sharing her vast knowledge of anatomy and her perspective on what is important to teachers; Doug Candland, for always letting me know when I had "lost my voice," and for his enthusiasm about the adventure of book publishing; and Bill Boswell, for the many hours he spent editing with extraordinary care, great skill, and humor.

The dear MTNA friends who provided personal support—Gail Berenson and Linda Cockey, who encouraged the writing of this book long before I entertained it as a real possibility and who provided continual support throughout the process.

My Bucknell University students, Lifelong Learning students, and all the people who have attended presentations at conferences and universities over the years. They not only gave me important feedback, but their enthusiasm made it clear that the impact of brain research on music study and performance isn't just my fascination but is shared by many.

Norm Hirschy, executive editor at Oxford, for his early enthusiasm for the book as well as for his endless patience in answering my many questions; Sean Decker and Rada Radojicic, Project Editors; cover designer Caroline McDonnell; the production team at Newgen, Koperundevi Pugazhenthi and Suma George; and copy editor extraordinaire, Timothy DeWerff. I am grateful for their care and expertise in making this book a reality.

The anonymous reviewers who made comments on an early draft of the manuscript. Their positive remarks let me know I was on the right track. Their observations and suggestions were invaluable in assisting me to explain complex ideas more clearly. I am extremely grateful to them.

Friends who are no longer here to see this book in publication but who had a significant impact on my ideas that would eventually become the book: Kay Payn, fellow explorer and dear friend, who was always eager to discuss ideas about the impact of brain research on teaching; Loren Amacher, gifted neurosurgeon, whose many comments about my early blog entries and his sharing of his experiences as a neurosurgeon provided not only useful insights, but also great humor; and Steve Mitchell, whose recovery from traumatic brain injury and return to performance was one of the early inspirations for the book, as were the many conversations in his instrument-crowded living room while we drummed on djembes.

My siblings, Anita Hellie, Renee Svard, Ron Svard, and Carl Svard, who put up with my years of piano practice at odd hours while we were growing up together, and who, with spouses Charles Hellie and Joan Svard, have been supportive cheerleaders throughout this entire project.

Nils and Katie Kresl, whose presence in my life is a source of great joy.

And most of all, my beloved husband Peter Karl Kresl, whose optimism, humor and unwavering support made this book possible.

About the Companion Website

www.oup.com/us/themusicalbrain

We have created a website to accompany *The Musical Brain: What Students, Teachers, and Performers Need to Know.* Links are provided to online interviews, TED talks, performances, demonstrations, articles, essays, and photos involving musicians or concepts discussed in the text. In addition, as newer research becomes available, links to that information will also be posted. The reader is encouraged to refer to the website to amplify or augment information in the text. Materials available online are indicated in the text with Oxford's symbol ⊙.

1

Music and Musicians—and Why the Brain Matters

... music heard so deeply
that it is not heard at all, but you are the music
while the music lasts.

—T. S. Eliot, *The Dry Salvages**

For centuries, poets and philosophers have written about the power of music, often suggesting that music is the essence of life itself, that music lives within us, that we *are* music. Scientists have dismissed these writings as flights of poetic fancy, or perhaps metaphor or artistic license. They have considered music to be a product of culture, and that's the way musicians have studied music as well. But have poets and philosophers perhaps had a better sense of the true nature of music? Have they been right all along in suggesting that music is life itself?

Most of us feel compelled to engage in music in some way, either by listening to or making music. According to a 2017 Nielsen poll, the average American listens to more than 32 hours of music a week.[1] That was up from 26.5 hours per week in 2016, and no doubt is much higher now following the coronavirus pandemic during which people were desperate to hear music, and many performers and organizations made performances available via streaming. Other than work or sleep, what other activity occupies so much of our time?

We listen to music for the emotion it communicates. Music can send chills or shivers down our spines. The soundtracks of movies and television heighten the emotional drama of the story. Music is an important part of every ritual, from commencements to weddings to funerals. We listen to

The Musical Brain. Lois Svard, Oxford University Press. © Oxford University Press 2023.
DOI: 10.1093/oso/9780197584170.003.0001

music when we're jogging, driving, walking, making dinner, and studying. And listening isn't the only way we experience music. Many of us sing in a choir or play an instrument in pick-up bands or community organizations, or we studied music as a child and sit down to play the piano or guitar from time to time. Some of us have become professional musicians, and music literally is our daily life. Our tastes in music vary widely—from classical to jazz, heavy metal, country, folk, rap, reggae, Indian ragas, or African drumming. But whatever our taste or level of engagement, music is never far from any of us.

All known societies in the world have some sort of cultural practice that could be described as music. Our compulsion to be engaged with music has ancient roots going back at least 40,000 years. Why is music so important to us? Although poets may have always written about music as part of life, scientists are beginning to change their previously held views that music is strictly a product of culture. Over the past thirty years, converging lines of inquiry and discovery in multiple scientific disciplines have led to a profound change in the way scientists think about music, and the evidence increasingly points to music as biological function. We may be as hardwired for music as we are for language—we may be born for music. The exploding field of neuroscience has paved the way for many of these discoveries.

We hear a lot about neuroscience, but what exactly is it? Neuroscience is not a single science; it is a multidisciplinary field that studies the structure and function of the brain and central nervous system. Neuroscientists come from a variety of backgrounds, including biology, medicine, neurology, computer science, chemistry, philosophy, linguistics, mathematics, and psychology. Neuroscientists basically study the impact of the brain on behavior and cognitive functions, or in other words, how does the brain produce the behaviors that we can see? Or, as we'll explore in this book, how does our brain wiring make it possible for us to learn an instrument and make music?

There has been an explosion of interest in neuroscience and the brain in recent years. It is not unusual to open a magazine and see beautiful colored images of the brain lighting up in response to pain, empathy, marijuana, or even anger. We see programs on television exploring how the brain encodes memories and emotions and how it processes conscious thought. Current books relating to neuroscience number in the tens of thousands, many of them becoming bestsellers. Titles tell us that if we change our brains, we can change our behavior, grades, love lives, become happier and smarter, and understand our emotions and habits. Neuroscience is now used to explain political decisions, marketing, and religion. We have become fascinated with the three-pound organ that sits atop our spinal cord and controls everything we do.

But despite our fascination with the brain, its allure often seems to be in the abstract, and we rarely think of it in connection with our personal lives. We certainly don't think about the brain in connection with music, or how we engage with music. We still think of music as something we do rather than as a part of who we are as human beings.

Music as culture

Musicians, as well as scientists, have traditionally looked at and studied music as a cultural product—an art form. We view works of music as stemming from a certain culture in a particular time period. The music of George Gershwin, for example, with its influences of jazz, Broadway, and Tin Pan Alley, could only have been written in early twentieth-century America. On the other hand, the Japanese drums called taiko were used in ancient Japan over 2,000 years ago. Known as the heartbeat of Japanese culture, the style of taiko used today was introduced to Japan by the Chinese and Koreans in the fifth or sixth century. The djembe, the most well-known African drum, dates from the thirteenth century in West Africa, and djembe drumming, with its tradition of multiple overlapping rhythmic layers, represents the multiple layers of life in West African culture. We place particular kinds and pieces of music within a certain time frame and a specific culture.

But scientists are now suggesting we look at music as something other than as a product of time and place. Very occasionally in the past century and a half, a scientist, while thinking of music itself as a cultural product, has at least made the connection between *making music* and the brain. In 1881, the German physiologist and physician Emil Du Bois–Reymond suggested that complicated muscle movements, such as those involved in gymnastics, fencing, dancing, and swimming, are controlled by the central nervous system—the brain and spinal cord.[2]

He wrote that it is inconceivable to think of the great pianists Franz Liszt and Anton Rubinstein without considering the "iron muscularity of arm," or to consider violinist Joseph Joachim without considering the force and speed of his bow arm. But he went on to say that the secret to the virtuosity of these great musicians was located in their central nervous systems, the roots of talent were in the gray substance of the brain, not in the hands. That gray substance refers to brain cells called neurons, and we'll see what they have to do with making music in Chapter 4.

In 1904, the great Spanish neuroscientist and pathologist Santiago Ramón y Cajal also referred to musicians when writing about the brain. In his *Textura del sistema nervioso del hombre y de los vertebrados* (Texture of the Nervous System of Man and the Vertebrates), Cajal wrote that a pianist acquires the skill through years of mental and physical practice, and this could be understood by acknowledging the strengthening of existing pathways in the brain, and the establishment of new ones.[3] He was speaking of neuroplasticity, or how the brain changes in response to experience and learning, even though neuroplasticity wasn't universally accepted in human adults until the 1980s. Neuroplasticity is what makes it possible to develop musical skills.

In 1967, a Russian pianist/teacher who had been living and teaching in New York since 1949 published a short, potentially groundbreaking book that discussed the role of the brain and spinal cord in making music. George Kochevitsky's *The Art of Piano Playing: A Scientific Approach* cites Du Bois–Reymond's work, provides a historical survey of various theories of piano technique, and then lays out his argument about the role of the central nervous system in piano playing. Kochevitsky writes that "practicing at the piano is mainly practicing of the central nervous system, whether we are aware of it or not." And "sooner or later piano pedagogy (as well as the pedagogy of any instrument) will have to accept the ideas in this book."[4] Unfortunately, his book was almost universally ignored.

Kochevitsky's work may have come nearly a century after that of Du Bois–Reymond, but he was still ahead of his time. Today, over fifty years later, musicians and music lovers still do not talk about, or even think about, the brain in connection with making or listening to music. On the other hand, over the past thirty years neuroscientists have thought about music and the brain a great deal, and they have found studying musicians and the process of making music to be a valuable way to study and learn more about human brain function.

Music as science

There are many cultures that don't use music notation, a visual system of symbols that represents musical sound, a set of visual instructions. But music notation is fairly standard in all Western music traditions. And while jazz improvisors usually perform without traditional music notation, the lead or fake sheets they use are a form of notation that indicates to the player all the essential elements of a tune, such as the basic harmonic structure and the melody.

Music notation is essentially a second language, and whether one is a beginner, an amateur, or a professional, the brain must process these symbols just like any other written language and translate them into motor commands in order to make sounds at an instrument or with the voice. At the same time, the auditory system is processing and monitoring the notes one is playing or singing to see if they correspond to the symbols on the page, and if not, fast corrections must be made.

Most instruments, whether keyboards, strings, winds, or percussion, require each hand to be doing something different, so precise bimanual coordination, also controlled by the brain, is very important. Add in the necessity of conveying emotion in the piece, planning the actions needed to play the piece, connecting what one sees with the spatial perception of how the body needs to move to accomplish all this, and perhaps memorizing and then recalling the piece, and it's no wonder that neuroscientists have found music to be such a rich area for study.

Robert Zatorre, cognitive neuroscientist and professor of neuroscience at the Montreal Neurological Institute, has written: "Indeed, from a psychologist's point of view, listening to and producing music involves a tantalizing mix of practically every human cognitive function."[5] And he goes on to say that even something as simple as humming a tune, which we all do, requires complex auditory processing, attention, memory storage and retrieval, motor programming, and sensory-motor integration. So one must surely ask, if even listening to or humming a tune is such a complex cognitive activity, why are we all able to do it? And how are we able to master the much more challenging skill of playing a musical instrument or singing at a professional level?

Since making music requires the coordination of sensory and motor processes and is so cognitively demanding, many neuroscientists have studied musicians to learn more about the organization of the human brain. And because long-term practice of a musical instrument causes both functional and structural changes in the brain, they can learn about neuroplasticity in all humans by studying musicians. The books that suggest you can change your brain in order to change your life are about neuroplasticity, the ability of the brain to change in response to experience or learning. Musicians have played an important role in helping neuroscientists learn more about how neuroplasticity functions in the human brain.

Brain networks for music

There is no single musical center in the brain, no "link" to Beethoven's "Für Elise" or your favorite pop song. Instead, there are multiple networks spread throughout the brain that process the different components of music. Over the past thirty years, neuroscientists have studied these networks, from those supporting the basic processing of pitch, rhythm, and melody, to more complex cognitive processes such as perception, sight-reading, emotion, improvisation, and memory. The neural substrates (circuitry) underlying some cognitive processes such as perception, memory, or emotion have been found to be the same as those used in our lives in general.

But other networks have been found to be used specifically for music. Think of "Happy Birthday." In the first "Happy birthday to you" the interval from "day" to "to" is a fourth. As you sing the second "Happy birthday to you," that interval increases to a fifth. And the third time, the interval from "Happy" to "birthday" stretches to an octave. Those kinds of pitch intervals, and the brain circuits that process them, are not used in language or for anything else in our lives. Language, even tonal languages in which pitch conveys meaning, such as Mandarin Chinese, have a very narrow range of pitches compared to music. So the brain networks used to process pitch intervals are unique to music.

Similarly, the complex rhythmic structures that we find in music have no equivalent in language, whether African drumming with its multiple rhythmic layers, Eastern European folk music with its alternation of accents in 3 and 2, or Paul Desmond's "Take Five," accented in groups of five and made famous by the Dave Brubeck Quartet. The brain circuits that process these complex rhythmic structures are as unique to music as are the circuits for pitch intervals. This suggests to researchers that music has functional or biological roots. And a great deal of recent research is focusing on the biological origins of music.

Imaging technology and music

Imaging technology, such as *functional magnetic resonance imaging* (fMRI), *positron emission tomography* (PET), and *electroencephalography* (EEG), has been instrumental in identifying areas in the brain that process musical elements such as pitch or rhythm. But it has also made it possible to look at questions that were, at one time, subject only to speculation.

In 2005, Lawrence Parsons, then at the University of Texas in San Antonio, scanned the brains of pianists in a PET scanner as they played the third movement of the Italian Concerto by J. S. Bach, and as they played major scales, all from memory.[6] While several brain areas were activated both in the playing of Bach and major scales, including the primary motor cortex and auditory areas, other areas supported either the concerto performance or the playing of scales, but not both.

Different kinds of jazz improvisation also activate different parts of the brain. In 2008, Charles Limb and Allen Braun, National Institutes of Health, scanned the brains of jazz musicians in an fMRI using a specially designed keyboard with no metal parts.[7] The jazz musicians improvised both to a C major scale and did a free improvisation. Brain scans showed a difference in activity in the two kinds of improvisation. Limb and Braun suggest that these findings may begin to provide a cognitive context for understanding creative activity in the brain. *[Limb presents this study in an entertaining TED talk that you can link to on the companion website, item 1.1 ⊙.]*

As we will see later in the book, our brains reflect our own individual auditory experience, so the brain of a classical pianist will look a bit different from that of a jazz pianist, a trumpet player, or a singer. The instrument we play, the kind of music we play or sing, and the amount of practice we engage in all have an impact on our brains. And it's not just motor and auditory activation that can be seen on a brain scan. How you process music emotionally can also be seen. The neurologist and best-selling author Oliver Sacks had his brain scanned at Columbia University Medical Center in 2009 while listening to music of Bach, which he said he loved, and Beethoven, which he did not. The music of Bach not only activated the many regions of the brain that have to do with the processing of the music itself, but also activated the amygdala, which is crucial to processing emotions. Beethoven, whose music Sacks was adamant about not liking, showed no activity in the amygdala. His brain did not react emotionally to Beethoven. Curiously, when an excerpt was played that Sacks did not know and could not identify as Bach or Beethoven, the amygdala in his brain lit up, knowing it was Bach.[8] *[A video in which Sacks and a researcher discuss this study can be seen on the companion website at item 1.2 ⊙.]*

Lesion studies and music

The oldest scientific technique for studying brain function, used long before imaging, and a technique still important and used frequently today, is to study individuals who suffer from some kind of neurological anomaly—brain-damaged individuals.

Isabelle Peretz, director of the International Laboratory for Brain, Music and Sound Research (BRAMS) at the University of Montreal, speaks of this patient-based approach as a kind of reverse engineering.[9] In healthy subjects, functional and neuroimaging studies have shown involvement of networks for processing music in both hemispheres of the brain in the temporal, frontal, parietal, and occipital lobes, and in sub-cortical networks (see Figure 4.3 in Chapter 4).

Studies of patients with lesions or other neural disorders can show more clearly which networks are directly related to music processing. Peretz suggests one can learn more about a complex system when it malfunctions than when it is running smoothly—reverse engineering. And neuroscientists and cognitive psychologists have learned a great deal about the neurobiological basis of music by studying individuals with various kinds of brain damage. This damage can be acquired or congenital and often leaves either language or music intact, but not both, suggesting that music and language are processed in music-specific or language-specific networks of the brain. Finding these music-specific networks lends support to the idea that music has biological roots.

Aphasia

Aphasia is a language impairment affecting one's ability to produce or comprehend speech, and sometimes the ability to read or write. It is most often caused by left-hemisphere stroke but can also be caused by brain tumors or a traumatic brain injury affecting the left hemisphere, primarily Broca's area, related to the production of speech, or Wernicke's area, responsible for the comprehension of speech. It can range from mild, having difficulty finding the names of objects—to severe, being totally unable to speak.

Aphasia does not affect intelligence, nor does it affect the ability to make music. In fact, therapists have known for some time that aphasic individuals can often sing sentences which they cannot speak. Neurologic Music Therapy uses elements of music such as rhythm, melody, and dynamics to treat aphasic individuals.

Neuroscientists often cite the Russian composer Vissarion Shebalin (1902–1963) as an example of an individual who was highly functioning in music even after losing his ability to speak. Although today we don't tend to know Shebalin or his music, he was appointed director of the Moscow Conservatory in 1942, and his music was well known at the time. In the late 1940s, Shebalin, Sergei Prokofiev, and Dmitri Shostakovich came under attack by the Communist Party of the Soviet Union for deviating from the party line in their training of young composers.

Shostakovich's and Prokofiev's struggles with the Soviet government have been well-documented; Shebalin's have not. But after several years of trying to defend his students and colleagues from government attack, Shebalin suffered a left-hemisphere stroke in 1953 that left his right side paralyzed and left him aphasic. (The left hemisphere controls the right side of the body; right hemisphere controls the left side, so a left-hemisphere stroke paralyzes the right side of the body.)

This was followed by another stroke in 1959. However, he continued to compose several works before his death, including two string quartets; a trio for piano, violin, and cello; a sonata for violin and viola; and several choral works—writing out all the scores with his left hand. He obviously was hearing the music in his mind, or he wouldn't have been able to notate it. Shebalin completed his Fifth Symphony just a few months before his death from a third stroke in 1963. Although this work has not become a part of the standard repertoire, Shostakovich called it "a brilliant creative work, filled with highest emotions, optimistic and full of life."[10]

A composer with whom we are more familiar, Maurice Ravel (1875–1937), is often cited as suffering from aphasia, but his is a more problematic diagnosis. The neurologist R. A. Henson wrote in 1988, and others have written more recently, about the difficulty in diagnosing Ravel's neurological issues. Although he had symptoms suggesting aphasia, neurologists have suggested frontotemporal dementia, Alzheimer's disease, and

Creutzfeldt-Jakob disease, and the jury is still out on his exact neurological disability. But nonetheless, even after he lost his ability to speak and to compose music, he retained his ability for auditory imagery and was able to hear music in his mind.[11]

Amusia

The reverse of aphasia is amusia, sometimes called tone deafness and described as an inability to process pitch accurately, whether perceiving or producing pitch. But the processing of musical rhythm, timbre, or emotion can also be affected. When rhythm is affected, it is called *beat deafness*, the inability to synchronize to a beat. But while music processing is affected, amusics can speak and understand speech perfectly well.

Amusia can be either congenital or acquired, usually as a result of stroke. In a ten-year study with a sample of 20,000 congenital amusics, Isabelle Peretz found that congenital amusia, present from birth, affects only 1.5 percent of the population and that in 46 percent of those cases, a first-degree relative also has amusia, suggesting a genetic basis. Acquired amusia as a result of stroke affects one to two thirds of stroke patients. Despite many people claiming to be "tone deaf" or to lack a sense of rhythm, the actual incidence of amusia in the general population is extremely small.[12]

In his book *Musicophilia*, Oliver Sacks relates his own experience with amusia. Sacks writes that while listening to a Chopin ballade on the radio, the pitch began to disintegrate and eventually sounded like "toneless banging." Rhythm was not affected, however, and he could still recognize the Ballade by the rhythm. A similar experience followed a few weeks later—this time when he was playing a Chopin mazurka on the piano. But this time, it was accompanied by zigzag patterns in half of his visual field. He realized he was experiencing a migraine aura. This strange reaction to music did not affect his speech or the sounds of others' speech—only music that he was listening to or himself playing.[13]

And in at least one case, fifth-grade teacher Margaret Haney developed amusia due to a viral encephalitis infection in one tiny portion of her brain. Having sung all her life, she found that she was suddenly unable to produce pitch accurately while she was trying to sing to her students. *[A link to a video of her amazing story can be found on the accompanying website, item 1.3 ⊙.]*

Brain areas causing amusia are quite different from those that cause aphasia, suggesting separate neural networks for language and music. Other kinds of brain anomalies also suggest separate wiring for music.

Musical alexia

In the opening chapter of *The Mind's Eye*, Oliver Sacks writes about the strange neurological condition of pianist Lilian Kallir, described by the pianist Gary Graffman as "one of the most naturally musical people I've ever known."[14] Kallir was born in Prague in 1931, demonstrated musical talent early, and gave her first public performance at the age of four. She was a sought-after performer, both as a soloist and as a duo-pianist with her husband Claude Frank.

Kallir, who was particularly known for her elegant performances of Mozart, had been scheduled to perform the Mozart Concerto No. 19 in F Major, but a last-minute program switch had changed it to the 21st Concerto in C Major. She had been playing all the Mozart Concerti for most of her life, and they were etched in her memory. But since it was a last-minute switch, she opened the score to check on a few things. But when she opened the score, it was completely unintelligible. She saw the lines, the individual notes, the rhythmic patterns, but nothing made any sense to her.

As the months went by, the problem persisted. If she was tired or feeling ill, she would be unable to understand musical notation, which she had known since the age of four. If she was feeling well-rested, it was fine. But eventually, she was unable to read musical notation at all, even though she continued to play beautifully and to teach, drawing from her prodigious memory. Some years later, she began to have problems reading words.

Alexia is a reading disorder, in which an individual is unable to read, even though an alexic may be able to spell and write. It usually occurs suddenly as a result of stroke or other brain injury. But according to Sacks, "Lilian was the first person I had encountered whose alexia manifested first with musical notation, a musical alexia."[15] Perhaps this is not surprising, since Kallir was probably reading music notation before she was reading her native Czech language and certainly before learning to read English. Musical notation was a native language to her.

Musical savants

Researchers have also looked at musical savants as evidence of the separation of music and language processing in the brain and a brain specialization for music. Savants are often found in autistic individuals. They may have extremely high levels of achievement in, say, music or chess, while being of below-average intelligence, with only basic linguistic skills. "Blind Tom" (Thomas Wiggins, 1849–1908) was an African American musical piano prodigy. He was one of the best-known performing pianists in America in the nineteenth century. He reportedly knew 7,000 pieces of music and composed some compositions, but he had extremely rudimentary speaking skills.

Derek Paravicini (b. 1979) is a British musical savant living in London. Born prematurely at twenty-five weeks, he grew up blind and with severe autism. But at the age of two, he discovered the piano and began teaching himself to play. By age nine, he was playing at the Barbican Hall with the Royal Philharmonic Pops Orchestra. Paravicini is a genius at improvisation. He can play almost any piece by ear and play it in any key or style. He has his own jazz quartet, a YouTube channel, has given a TED talk, has a packed concert schedule, and has appeared on CBS's *60 Minutes*. Yet he is linguistically challenged and has trouble communicating with language, but not with music. *[A link to his TED talk is found at item 1.4 on the companion website ⊙.]*

Why musicians should know about neuroscience

We are fascinated by individuals such as autistic savant Derek Paravicini, who enjoy a rich musical life despite great odds. We feel sympathy for someone like Margaret Haney, a grade-school teacher who had enjoyed singing all her life and suddenly could no longer find the right pitches. But we wonder, what does this have to do with us?

We may not be able to relate to those particular instances, but they are a part of the larger neurological puzzle of how and why we are able to make music. Imaging and lesion studies have demonstrated that there are some separate brain areas for music and language. As we will see in the coming chapters, other kinds of investigations have uncovered a great deal of information about why music exists and how and why we are able to make music. Two themes run throughout this book: (1) music has biological foundations—humans are hardwired for music just as they are for language;

and (2) neuroplasticity, the extraordinary ability of the brain to change in response to learning or experience, makes it possible for us to develop to an exceptional degree the musical abilities with which we are born.

You may be thinking: "musicians have performed beautifully, often spectacularly, over the past couple of hundred years without knowing anything at all about the brain, and they will continue to do so. So why do we now need to know about the brain and music?" This may be true, but what is fascinating about neuroscience and music research is that it shows us that music is not just "something we do." Music is fundamental to who we are as human beings. Our prehistoric ancestors were making music tens of thousands of years ago, and our brains have evolved not only to support music-making but to value it. Researchers have discovered a great deal about the brain and the process of making music that can help us learn more efficiently, teach with a better understanding of "how musicians learn," and perform with far greater confidence. Learning about the brain adds a fascinating dimension to what we already know—or think we know—about making music and its place in our lives.

Key concepts

- Music has traditionally been studied as a cultural product.
- Since making music requires complex auditory processing, memory storage and retrieval, motor programming, and sensory-motor integration, scientists are increasingly studying music as biology to learn more about the structure and function of the human brain.
- There is no single area in the brain where music is processed. Instead, different elements of music are processed in many areas throughout the brain and vast neural networks connect these areas.
- Imaging technology and lesion studies have both been instrumental in identifying brain areas that are involved in processing music.
- Research into conditions such as aphasia, amusia, musical alexia, and musical savants suggests that music and language are processed separately in the brain.
- Discoveries in neuroscience, behavioral and cognitive psychology, archaeology, biology, and other areas are leading researchers to conclude that music has biological foundations, and that we are hardwired for music as we are for language.

2

Origins of Music

A key part of being human is being musical.
—archaeologist Stephen Mithen[*]

On a balmy May evening several years ago, my husband and I were walking along a downtown Montreal street on our way to hear a Montreal Symphony concert. Somewhere ahead, we heard the clanging of pitches and rhythms, not unlike the sound of tubular bells. Maybe an outdoor concert? But as the pitches grew louder, we found that we were approaching what appeared to be a building in the process of demolition. We looked up three or four floors toward the fragments of melodies we were hearing. The outer side wall of the building had been removed, exposing steel support beams and an assortment of pipes. Three men were taking a break from their deconstruction work and were striking steel beams and pipes as though they were instruments. The men were having a great deal of fun using short pieces of pipe and other metal tools as mallets to create short tunes. Laughter and shouting along with the clanging pitches drifted down to the street below, mixing with the delighted clapping of the crowd that had gathered.

What is music? One of the difficulties in studying the origins of music has been the definition of music itself. Every researcher who studies music, whether coming from the broad fields of music or science, defines music in a slightly different way, considering cultural differences, historical time periods, and even emotional differences. I like the definition proposed by

The Musical Brain. Lois Svard, Oxford University Press. © Oxford University Press 2023.
DOI: 10.1093/oso/9780197584170.003.0002

composer Anthony Brandt, who defines music as "creative play with sound; it arises when sound meets human imagination."[1] That definition would certainly apply to the sounds created by the construction workers, and it would also pertain to the self-taught percussionists playing on street corners in every city during the summer, children beating on whatever pots or pans are available, or adults pinging the crystal glassware at the dinner table. "Sound meeting imagination" to make music seems to be in our DNA.

Oldest musical instruments

In the summer of 2008, archaeologist Nicholas Conard from the University of Tübingen led excavations at Hohle Fels in southwestern Germany that unearthed one nearly complete bone flute and fragments of two ivory flutes carved from mammoth tusks. The flutes have been radiocarbon-dated to about 40,000 years ago during the Upper Paleolithic Era, which extended from about 50,000 to 10,000 years ago.

Our prehistoric ancestors are often portrayed in cartoons or on television as being quite primitive, but both the musical sensibilities and the technological sophistication of the flute makers 40,000 years ago are astonishing. The Hohle Fels bone flute was made from the hollow wing bone of a griffon vulture. The bone had been smoothed by scraping, and V-shaped notches were cut at the top to form the mouthpiece. A series of fine lines perpendicular to the length had been marked, either for ornamentation or to measure the position of the five finger holes.[2] The Hohle Fels flute is shown in Figure 2.1.

An earlier dig at a nearby site at Gießenklösterle, also led by Conard, uncovered fragments of two smaller three-holed bone flutes made from the radius of a whooper swan and fragments of a flute made from mammoth ivory. All six flutes, the three from Hohle Fels and three from Gießenklösterle, were originally dated to about 35,000 years ago.[3] More recent dating in 2012 by the Oxford Radiocarbon Laboratory has pushed that date back to between 39,000 and 43,000 years ago, about the time anatomically modern humans were spreading into Central Europe along the Danube River Valley.[4]

The ivory flutes from both sites demonstrate even more sophisticated technological thinking than the bone flutes. The flute makers carved a straighter tube from the larger curved mammoth tusk, split it, hollowed out the two halves, and carved evenly spaced finger holes. Little notches were cut so the halves would fit together and some kind of sealant glued the two halves

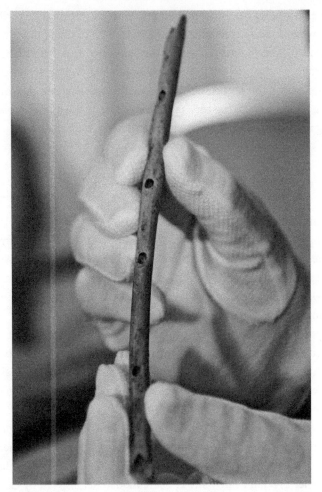

Figure 2.1 Archaeologist Nicholas Conard showing the Hohle Fels bone flute, dated between 39,000–43,000 years before the present.
Credit: © AP Photo/Daniel Maurer

together. The process would have required a significant amount of creativity and technological expertise.[5]

At Isturitz, one of the most important Paleolithic sites in France, over two dozen flutes made from the bones of large birds have been found and dated to between 20,000 and 35,000 years ago. At several other sites, archaeologists have found objects, or parts of them, that are classified as pipes, flutes, or whistles. They date to the Upper Paleolithic Era, from 12,000 to 32,000 years ago. Iain Morley's *The Prehistory of Music* details the more than 140 flutes or pipes that have been found at Paleolithic sites.[6]

What is music?

Finding one or two flutes from tens of thousands of years ago would be remarkable. But finding over 140 of them means that flute playing was not an isolated activity. Making music was an important part of the culture. Humans obviously come from a long line of music-makers. Perhaps clanging on pipes or pinging on water glasses is part of our biological heritage.

Charles Darwin (1809–1882), the English naturalist, geologist, and biologist, was the first scientist to suggest that music was evolutionary in origin, that it is part of who we are. In *The Descent of Man* (1871), Darwin wrote that music arouses such great emotions in humans because it once served the same function in humans as it does in birds, finding a mate, and sexual selection affected likelihood of reproduction and the maintaining of the species, even though that function no longer exists. He allowed that "neither the perception nor the production of music were faculties of the least use to man."[7] Nonetheless, he believed that because music had originated in vocal expressions of emotion and it had a function in sexual selection, it evolved as a human capacity that was a precursor to the capacity for language.

On the other hand, only twenty years after Darwin wrote about the biological origins of music, William James (1842–1910), the father of American psychology, wrote in his *Principles of Psychology* (1890) that "musical sounds are without any utility whatever," and that love of music was "a mere incidental peculiarity of the nervous system." Rather than evolving from a primal response, as Darwin had suggested, James wrote that music "entered the mind by the back stairs."[8] In the rare instances when scientists addressed music during the next century, it was from the standpoint of one of these opposing views: either it was evolutionary, or it was incidental. But beginning in the late 1980s, the theory of evolution began to be considered as a potentially valuable way to think about certain aspects of music, and many researchers began to study music's origins.

In 1997, a century after James's comments about music but extending his line of thought, cognitive psychologist Steven Pinker declared music to be "auditory cheesecake," pleasurable, but of no real value. Pinker had been asked to give the keynote address at a conference of the Society for Music Perception and Cognition (SMPC), held that year at the Massachusetts Institute of Technology. Pinker was known at that time for his work in language and the mind, and while he considered language to be an evolutionary adaptation, he thought music was a by-product. According to Pinker, we like cheesecake, or music, because it activates our pleasure circuits, not because of any intrinsic value—"empty calories," so to speak.[9] One must wonder if music is of no value, why did prehistoric humans spend so much time crafting flutes, and why has music persisted for so many tens of thousands of years?

Pinker's comments weren't addressed to a neutral audience. All the researchers in the room had spent years studying various aspects of music, and not surprisingly, there was a great deal of anger, disbelief, and consternation among the assembled crowd. Pinker had even stronger words about music in his book *How the Mind Works*, published later that year. "As far as biological cause and effect are concerned, music is useless. . . . Music could vanish from our species and the rest of our lifestyle would be virtually unchanged."[10]

Would your lifestyle be unchanged without music? Whether one makes music professionally or just enjoys listening to it, life without music is unimaginable for most of us. Based on archaeological finds such as the Hohle Fels or Gießenklösterle flutes, life has been unimaginable without music for at least 40,000 years, and likely for tens of thousands of years before that. If music were really of no use, surely it would have vanished from our species thousands of years ago.

At the time Pinker made his comments, research in the origins of music and its cognitive underpinnings was already ongoing, not only by the many assembled at the conference, but by researchers around the world. His comments, however, brought new focus and urgency to the work, and researchers began to address such questions as: What is music? What does it mean? Where did it come from? Why does every human culture have it?

Music vs. musicality

Music and language are universal human traits, and they have always been compared when biological origins are discussed. But while language has the definite purpose of communicating facts, ideas, attitudes, and abstract thought—it signifies or means something—music doesn't convey meaning, facts, or ideas—it conveys emotion and feeling. So the question for researchers has been, Why do we have music? One of the stumbling blocks in defining music's biological foundation has been the term "music" itself and how it is defined. While every culture at every period in history has had some practice that could be called "music," what is considered music in one culture may be barely recognizable as such in another. Even though globalization has caused a significant cross-pollination of musical cultures, there are still indigenous peoples in various parts of the world who do not recognize what we listen to as "music." In some languages, there is no specific word for music—the term encompasses rhythm and dance as well.

Early researchers delving into the origins of music essentially said, "Everyone learns language effortlessly, we absorb our native language as we grow up, but music requires lots of years of practice and study, so it can't be biologically based." This compares an average adult's language capabilities to that of the general population but compares the same adult's musical abilities to those of professional musicians and virtuosi—the proverbial comparison of apples to oranges.

Steven Mithen pointed out that a "natural biologically based musicality" and actual music are not the same thing.[11] Other researchers began to realize that if they were going to talk about music as biology, they needed to distinguish between *music* as a cultural product and a more generalized idea of *music* referring to the brain processes that underlie musical behavior.[12] *Music* is now used in origins-of-music research to refer to the social and cultural construct that reflects the time and place in which it was created—the actual creative works that we compose, listen to, learn, and perform. The term *musicality* is used to refer to a naturally developing set of traits, the brain processes that have evolved over time that underlie musical behavior, such as the ability to pick up a beat, to recognize a tune no matter what pitch it starts on, and to be able to sing with a group, all of which most adults can do.

Unfortunately, this is one of the instances, and there are others, in which scientists and musicians use a term in different ways, creating confusion. Scientists in various disciplines don't necessarily even use the same terminology. Researchers studying music as biological have adopted the term *musicality* to refer to the brain processes supporting musical behavior, behaviors that may have an evolutionary basis. But for a musician, *musicality* is the ill-defined quality of bringing expressiveness to a performance that is often based on a printed page, of creating a beautiful sound, of interpreting a composer's intentions, of being able to communicate the emotional essence of the work. A musician recognizes that one can perceive and make music without being musical—having a good command of technique without being able to communicate any emotion. Yet to scientists in this field, all people possess musicality simply by virtue of being human.

As Isabelle Peretz and others have pointed out, the average adult who has had no musical training still has quite a few musical abilities.[13] Nearly everyone can carry a tune, even without musical training. Most people will join in singing "Happy Birthday" or their country's national anthem—if they are pitched reasonably. Nearly everyone can recognize a familiar melody and can recognize if an odd note is inserted into that melody, no matter what pitch it begins on. Nearly everyone can keep a beat—we dance in rhythm, we clap in rhythm. A lot of people pick up the ability to play an instrument without formal study. Most adults can distinguish the difference in periods of classical music, even if they don't know what they are called.[14] For example, they can recognize Bach's music from the early eighteenth century as being from a much earlier era than the music of Aaron Copland, a twentieth-century composer. And most will decry today's pop music as much worse than what they listened to growing up, clearly recognizing the difference in style. We take those musical abilities for granted.

According to Peretz, "Musical abilities are widely distributed in the population, probably on a continuum of musicianship with poor abilities at one extreme and superior abilities at the other. The vast majority lies in the middle with a common core of musical knowledge but modest production skills."[15] Most people have a basic knowledge of music even if they aren't skilled at making music themselves.

Those musical behaviors are as natural as language. It takes years of practice and study to become a skilled musician, but almost everyone has basic musical skills that allow enjoyment of music and even music-making at some level. Music teachers know it is possible for someone with limited performance skills to perform musically, just as it is possible for someone with great technique to have little musicality. Everyone may learn his native language effortlessly, but it also takes years of practice and study to become a skilled writer, public speaker, or actor.

Evolutionary theories of music

In the past twenty to thirty years, there has been an explosion in research concerning music perception and performance and their correlates in the human brain. Researchers have looked at music to better understand the organization of the brain, and to find out more about the "functional origin and the biological value of music."[16] Remarkable discoveries have been made in the neurosciences, as well as in fields such as neurology, anthropology, archaeology, behavioral and cognitive psychology, music theory, ethnomusicology, audiology, and linguistics. A consensus has been building that music indeed has biological or evolutionary origins.

There are three prominent theories about music as having adaptive or evolutionary origins (which are dealt with in more detail later). The first is Darwin's view, that of music's early function in sexual selection. A second theory for music as adaptive or evolutionary concerns mother-infant communication or bonding. The third theory, which has achieved the most consensus among researchers, expands the focus of communication from bilateral between mother and infant to communication among members of a group. A common thread running throughout the discussion of music as biology is the question of language versus music—which came first? Steven Pinker and others who believe music originated as a cultural invention believe that language evolved first and music was an invention that tapped into, or piggybacked on, brain functions supporting language. The several theories that support music as evolutionary or adaptive argue that music evolved before language.

Language and music

The origins of language have been studied far longer and more extensively than the origins of music. The basic question has been, How did the human species evolve to become capable of developing this complex system of communication? Most of the research being done today on origins of vocal production or sound centers on language, when and how did our prehistoric ancestors develop language. But researchers who study the origins of music have concluded that a kind of music existed long before language.

While Darwin was the first to suggest that music has biological origins, Mithen is perhaps the most well-known today of the adaptionists, those who believe that music has a biological, evolutionary basis. Mithen, cognitive archaeologist at the University of Reading in the UK, has been a strong proponent of the belief that the origins of sound or vocal production stem from music. In his fascinating book, *The Singing Neanderthals: The Origins of Music, Language, Mind, and Body*, he lays out a compelling case for music as fundamental to humanity, encoded in our genome through evolution.[17]

Mithen believes, as did Darwin, that music was essential for the survival of our Stone Age ancestors, and while it is no longer essential for survival, we have inherited our compulsion to engage with music, whether by listening to or actively making it. He suggests that the roots of music go back to the Neanderthals, who lived from about 400,000 years ago to 40,000 years ago in Europe as well as southwestern to central Asia. He proposes a proto-musical language based on variations in pitch, rhythm, and timbre that, in addition to its importance in sexual selection, was also important for mother-infant bonding and bonding within a group.

The view of Mithen and others who promote the mother-infant communication idea is that with the long period of infantile human dependence when a mother needed both hands for making food and other tasks, she maintained connection to her infant through sound, through the proto-musical language.[18] Even today, babies respond more strongly to singing than to speech.

Most researchers who believe that music has biological origins believe that music evolved because it helped to promote group cohesion through *emotional contagion*.[19] Emotional contagion occurs when one person's actions and emotions trigger similar actions and emotions in others. Getting caught up in the emotions of those around you at a rock concert or political rally is emotional contagion.

Without possessing a spoken language, this early proto-musical language was an important way for early humans to survive—to communicate about hunting for food and developing the capacity for big-game hunting, to communicate emotion, to cope with dramatic environmental changes, and to withstand pressures or assaults from other groups.[20] This proto-musical language eventually split into language and music, language to express abstract thought and communicate ideas and facts, and music carrying emotional content more exclusively. It is pitch and rhythm that make emotional contagion and group bonding possible, and neither occurs in speech. We don't synchronize to language, but we do to music. (Chanting at athletic events, political rallies, and protests may involve language, but it is also considered by many to be a form of music.)

Experiencing music in a group is far more prevalent today than solo music-making, and it still provides a bonding function in the singing of church hymns, school songs, national anthems, pop songs, and concerts of all kinds where people share an emotional experience through music.

The first musical instrument—the voice

Forty-thousand-year-old flutes notwithstanding, researchers uniformly believe that the first musical instrument was the voice itself, in whatever fashion Neanderthals were using it. It requires nothing external; sound production is within the body. Although Mithen's premise is that Neanderthals used a proto-musical language, he does not cite any direct evidence for singing Neanderthals, and with good reason. A major problem determining the evolution of vocal production, whether for language or music, is the lack of fossil evidence for speaking or singing. Most of the physiology that is responsible for our ability to make sound is cartilage and soft tissue, and neither survives for tens of thousands of years. But archaeologists have drawn conclusions about how sound production may have worked in Neanderthals by examining the bony structures that support sound production, and bones have survived in the fossil record.

In 1983, a skeleton was excavated from Kebara Cave in Israel. Called Kebara 2 (Kebara 1 was an incomplete infant skeleton), it has been dated to about 60,000 years ago, which means it was Neanderthal. The skeleton was mostly intact, but what was of particular interest is that the Kebara skeleton contained the first intact hyoid bone found in any pre-modern human.[21]

If you have watched any crime dramas on television, you are probably familiar with the hyoid bone, since investigators in these dramas often refer to a broken hyoid as evidence of strangulation. The hyoid is a small bone at the front of the neck, roughly in the shape of a U, that serves as an anchoring structure of the tongue and a support for the larynx. It is attached to other bones only by soft tissue. Since it supports the tongue, it is crucial for speaking and singing.

Not only was the hyoid intact in Kebara 2, but it has also been found to be nearly identical in form to that of modern humans. Using 3D X-ray imaging and bone mechanical modeling, an international team of researchers found that this Neanderthal hyoid bone not only looked like a modern hyoid but was used in a very similar way.[22] Other anatomical evidence has demonstrated that Neanderthals appeared to have motor control over their tongues and breathing that was similar to modern humans, meaning that Neanderthals had vocal capabilities equivalent to those of modern humans, although the range of sounds was no doubt different.[23] There is no fossil evidence to show that Neanderthals sang or spoke, but there is fossil evidence to show that they had the physical capacity to do so.

Analysis of bones in the inner ear in other skeletal finds has shown that by 300,000 years ago, it was possible that sound perception was equivalent to ours today.[24] Neanderthals were capable of hearing the same frequencies as humans can hear. Mithen suggests that, in order to have survived for hundreds of thousands of years through major climate change, Neanderthals must have been able to cooperate to hunt and gather food and to maintain social relationships. Some kind of communication would have been necessary, and there is no evidence that language developed that early. Mithen's argument for a proto-musical language is compelling.

Rhythm and drums

Most people can keep a beat; we synchronize by nodding our heads, clapping, tapping our feet, or dancing. There are various theories about how that ability evolved. Early man may have responded to his heartbeat, the most basic rhythm of all. Several researchers speak to bipedalism as the impetus for rhythm.[25] We needed to develop an internal rhythmic mechanism to be able to walk, and that internal rhythm has been with us ever since. Even though we don't think about it, we walk in a rhythm. Researchers have known for some time that people who suffer from Parkinson's and have difficulty in walking respond well to music or even to just a rhythmic beat. Their gait improves when they walk to music that they like. The external beat provides the rhythm that their internal mechanism is not providing.

Some researchers speculate that rhythm may have originated when our ancestors were pounding rocks to extract grains and discovered they could synchronize, or when they clapped their hands together and discovered they could get a louder sound by striking rocks.[26] So perhaps it is not a stretch to believe that early humans translated those rhythms into striking a stone or wood surface—early drums. Since producing sound of any kind involves using the body, whether the voice to sing, the mouth to blow on a flute, or arms and hands to drum, making music or a rhythm is *embodied*. Our bodies are producing the sound and we feel the music within our bodies.

Other Paleolithic instruments

There is no way to tell if a specific piece of rock or stone was used for drumming, and any drums that may have been fashioned from wood and animal skins would have decomposed long ago. But one rather amazing Paleolithic percussion instrument does survive.

At the site of a Paleolithic settlement near Mezin, Ukraine, dated to 20,000 years ago, archaeologists found the remains of a house built from the bones of a mammoth. Inside this probably communal hut, several mammoth bones were discovered each containing a cut-out geometrical design colored red—a set of bones. In the same place, they found two ivory rattles, a mallet made from a reindeer antler, and a bracelet made of five pieces of mammoth tusk ivory, with carved decoration. Nearby they found pure yellow and red mineral ochre that had apparently been used for painting the bones.[27] [A link to photos of these instruments can be found on the companion website, item 2.1 ⊙.]

The original dig took place between 1954 and 1962, but it wasn't until 1974 after several years of study that a team of archaeologists, paleontologists, forensic scientists, and medical experts concluded that these several bones, the mallet, and rattle formed a set of percussion instruments, perhaps the first percussion orchestra. The decorated bones included a shoulder blade, thigh bone, jaw bones, and skull fragments. The pattern of wear showed that they had been struck repeatedly, and each of the bones produced a different pitch. The rings of the ivory tusk bracelet made a rattling sound when shaken, like a castanet, leading the researchers to conclude that dancing was a part of the music-making.[28] Making music was important to these early humans and they used what they had at hand, mammoth bones.

Using steel beams and metal pipes to produce pitches and rhythms in the early twenty-first century is not unlike using mammoth bones to create pitches and rhythms in 20,000 BCE. Both stem from the same musical impulse, to create sound and rhythm using whatever materials are at hand. In addition to the voice, early flutes, drums, and an early percussion orchestra, other instruments have been identified. Archaeologists have discovered phalangeal whistles, usually made from the first or second bone of a reindeer's foreleg; bullroarers, made from a flat piece of wood, stone, bone, antler, or ivory attached through a perforated hole to a cord and spun in a circular motion over one's head, making a whirring sound; rasps, similar to a modern guiro; and rattles, two ivory rattles having been found at the mammoth bone orchestra site and speculated to be used in dancing. Even environmental surfaces, such as caves, were used to create sound. Prehistoric humans were extremely inventive in creating musical instruments. *[Links to photos of several Paleolithic instruments can be found on the companion website, item 2.2 ⦿.]*

Cave art and music

Bone and ivory flutes, bullroarers, and small figurines have often been found in or near Paleolithic caves, and there is probably a reason for that. Archaeologists consider the caves themselves to be a musical instrument. A cave can be a giant resonator, amplifying footsteps, voices, dripping water, or sounds made by water running beneath the cavern. Stalactites and stalagmites ring like giant tubular bells when struck, and even water dripping onto a stalagmite can "ping."

Acoustics in a cave can vary a great deal depending on the size of the space, but also on the composition of the surfaces, whether limestone, chalk, gypsum, or some other kind of rock. Over the past thirty years, studies have shown a relationship between cave paintings and the acoustic properties of the area of the cavern in which they are found, indicating that early humans sought locations for art that were also resonant for sound or music.

Iégor Reznikoff, University of Paris, has a PhD in mathematics and describes himself as a specialist in the resonance of buildings and spaces, but he has primarily been associated with the resonance of Romanesque churches. He is also a pioneer in the field of archaeoacoustics, the archaeology of sound. As singers often do, he sings or hums to himself whenever he enters a new space to learn about the resonance in that space.

Not surprisingly, in 1983 when he first visited a French cave with prehistoric art, he sang and hummed in different parts of the cave to test the resonance, just as he was accustomed to doing in churches. He discovered that the sounds were louder and more resonant when he was in areas with painted animals, and he wondered if there might be a connection between the *amount of resonance* and the *locations* of the paintings in a cave.

To test his idea, Reznikoff visited more than ten painted caves in France with art ranging from 15,000 to 25,000 years old, including Niaux and Le Portel in Ariège and Arcy-sur-Cure in Burgundy, singing as he explored. He found that more than 90 percent of the paintings were located in the most resonant parts of a cave where echoes would reverberate for some time. The density of paintings in a location was proportional to the intensity of resonance. In areas such as narrow passages where painting would have been difficult, but the acoustics were still good, there were often markings of red dots.[29]

More recent studies confirm these findings. In 2013, Rupert Till, an acoustic archaeologist at the University of Huddersfield, UK, and a group of researchers explored five of the Cantabrian Caves in Spain.[30] They used a laptop and speaker to sweep a sine wave tone through all audio frequencies to get an acoustic fingerprint of each space in the caves.[31] They found that the smaller, less acoustically resonant spaces contain the older paintings, up to 40,000 years old, and these images tend to be simple dots or handprints. The larger, more resonant spaces contain paintings of animals—deer, bison, horses, mammoths—and these date from 15,000 to 25,000 years ago. The spaces with the animals are large enough for groups of people to have gathered for rituals. When hearing someone play a reconstruction of a 40,000-year-old bone flute in complete darkness, Till remarked that "the music seemed to bring the environment to life."[32]

Both Till and Reznikoff believe the correlation of cave paintings with the most resonant chambers in a cave suggests a ritualistic or religious reason for the paintings. Reznikoff says, "Indeed rituals and celebrations are mainly based on singing and music, and why would the Paleolithic tribes choose preferable resonant locations for painting if it were not for making sounds and singing in some kind of ritual celebrations related with the pictures?"[33] *[On the companion website, you will find a link at item 2.3 to a website that Till and his colleagues created attempting to reconstruct what it may have looked, sounded, and felt like to be in a cave in prehistoric times. You can hear clips of Reznikoff singing in the cave of Arcy-sur-Cure at item 2.4 ⊙.]*

Before 40,000-year-old flutes

Early humans didn't arrive in Europe 40,000 years ago and suddenly have the expertise to make complex flutes with sophisticated tuning systems. To not only possess the technological knowledge to construct bone or ivory flutes but also to have a somewhat sophisticated concept of pitch implies a period of long evolutionary development and previous experience with music and flute-making.

Flutist and biologist Jelle Atema, who has made reconstructions of several Stone Age flutes, describes a possible series of developmental stages in flute construction that begins with a hollow tube (reed or bamboo) capable of one pitch, eventual addition of tubes of varying lengths to produce several pitches (pan pipes), adding finger holes to the pipe to produce more pitches, and eventually leading to the transverse flute we know today.[34] This would have happened over a period of thousands to tens of thousands of years. The flutes found at Paleolithic sites in Germany and France were already well along his proposed developmental timeline of historical flute technology.

These innovations in flute making would have been driven by both technological and musical curiosity. How could one create more pitches, extend the range, make a bigger sound, make the flute sound softer or brighter, convey emotion more clearly? Innovations would only have happened if music had been extremely important in the lives of these early humans.

A flute has been discovered that is considerably older than the Paleolithic flutes, but it is shrouded in controversy. The Divje Babe flute was found in a Slovenian cave in 1995. This was a Neanderthal site, and the flute has been dated to between 50,000 and 60,000 years ago.[35] The flute, made from the femur of a one--to-two-year-old cave bear, is about 4 ¾ in. long with a U-shaped notch at one end and has two complete round finger holes and one broken hole. The Divje Babe flute is shown in Figure 2.2.

The find is controversial because it has been a long-held belief that Neanderthals did not have the capability of abstract or symbolic thought. So how could they have produced an instrument?

The scientists who contend that it could not be a musical instrument believe the holes were made by scavengers such as hyenas or wolves looking for bone marrow. However, zoologists contend that scavengers would have crushed the bone, not made three neat holes in a row, holes that perfectly fit human fingers.[36]

Science is rarely without controversy. For a theory to become recognized as fact, years of research to confirm or refute existing and newly accumulated evidence is required. Newer technology is constantly being developed that may evaluate previous evidence in a new light. At one time, it seemed that Neanderthals were cognitively far less advanced than the early humans who arrived in Europe about 45,000 years ago. This theory was based on lack of evidence of Neanderthal fossils, objects, or artifacts. But as scientists say,

Figure 2.2 The flute from Divje Babe, 60,000 years before the present
Credit: © National Museum of Slovenia, photo Tomaž Lauko

"Absence of evidence is not evidence of absence": just because evidence hasn't been found does not mean it doesn't exist.

Over the past twenty years, several finds have shown that Neanderthals were far more sophisticated than we thought. They knew how to make twisted fiber or cord 50,000 years ago. They produced cave art in Spain over 64,000 years ago, which is much earlier than the earliest *Homo sapiens* cave art of 32,000 years ago. They made bead shells that have been dated to 115,000 years ago. And they made a birch bar tar for use as an adhesive 200,000 years ago.[37] By the time flute makers were using a sealant to glue together the two halves of a mammoth ivory flute 40,000 years ago, sealant or glue had been around for tens of thousands of years, developed by Neanderthals.

All these discoveries, and there will no doubt be more, show that Neanderthals were far more cognitively advanced than previously supposed. It seems reasonable to conclude that if they were capable of the objects and discoveries just discussed, they could also have made a flute. Flute makers at Hohle Fels and Gießenklösterle didn't suddenly have the knowledge to make flutes out of bird bone and mammoth ivory without a long history of experimentation that had to have happened somewhere. Whether or not the

Divje Babe artifact itself falls in that timeline may be open to question, but Neanderthals appear to have been technically advanced enough to make one. *[A short film including a performance on a replica of the Divje Babe flute can be heard on the companion website at item 2.5 ⊙.]*

How did early musical instruments sound?

We don't know how our prehistoric ancestors used their voices, nor do we know what kinds of rhythms they may have made while drumming. We do have an idea about how their flutes sounded because it has been possible to make replicas. Experimental archaeology is a sub-field of archaeological research in which archaeologists try to understand prehistoric people by replicating objects that have been found at archaeological sites using the same tools used to make them originally. Wulf Hein is an experimental archaeologist who specializes in the reconstruction of various kinds of prehistoric finds, from making the tools themselves to carving small figures or constructing replicas of prehistoric flutes. He speaks about and plays the replica he made of the Hohle Fels flute in Werner Herzog's 2010 documentary, *Cave of Forgotten Dreams*, about France's Chauvet Cave (although the flute was not found there).[38] *[A clip can be found on the companion website, item 2.6; Two other performances on reconstructions of the Gießenklösterle bone flute and mammoth ivory flute can be heard on the accompanying website, item 2.7 and item 2.8 ⊙.]*

Archaeologist Hein says that reconstructing prehistoric cultural artifacts has taught him something important about the culture.[39] Using the same techniques and kinds of tools that prehistoric sculptors would have used, Hein recreated a figure called the Lion Man, a 40,000-year-old ivory sculpture, initial fragments of which were discovered in 1939 in the Hohlenstein-Stadel, a Paleolithic cave in southwestern Germany. Additional fragments were found in the 1960s and early 1970s and pieced together in 1982. The figure is about 12 inches tall and a little over 2 inches around. As he worked on the replica, Hein counted his work hours.

He stopped counting after 400 hours, so it took longer to complete the figure. But while sculpting, he realized something significant about that early culture. Life in prehistoric societies revolved around the work of providing food and shelter; community members spent their days hunting, fishing, or gathering edible roots and berries. Hein suggests that someone was freed from those tasks to use the time to fashion this incredible ivory figure. Similarly, others would have been given time off from the work of the community to make the bone or ivory flutes. Constructing finger holes on a flute demonstrates knowledge of pitch, and constructing flutes of different materials shows a consideration of sound qualities because bone and ivory flutes have different timbres, so there must have been a tradition of individuals learning about the properties of flutes by making them. The amount of time, creativity, and technological expertise necessary to create art or flutes shows that art and music were important in this early culture.

Why should we care about our prehistoric ancestors making music?

In his book *Descartes' Error* (1994), neuroscientist Antonio Damasio argues for the importance of emotions in rational thinking and decision making.[40] Mithen suggests that "being emotional is essential to being intelligent, making effective decisions, and being a successful member of a social group."[41] Without language to express thoughts and ideas, communicating emotion was necessary for survival, and the earliest humans communicated emotion by using a proto-musical language, with variations in pitch, rhythm, and timbre. We have inherited the need for language to express our thoughts and ideas. We have inherited the need for music to express or to feel emotion. Yet, many people, such as Steven Pinker, dismiss music as being a frill—"auditory cheesecake." Why should we dismiss the music portion of our biological inheritance as any less valid than language?

Archaeological discoveries provide ample evidence that music in some form has been a part of the human experience for tens of thousands of years. Our prehistoric ancestors could not have survived without music, and neither can we. The changes in our brains, both structural and functional (as we'll see in Chapter 4), that result from making music have a positive impact on many areas of our lives, whether we study for two years or the many years required to become a professional musician. We enter the world with biological predispositions to music and with musical abilities. Why not develop these musical abilities the same way we develop our language abilities, to make us more fully human?

Key Concepts

- Earliest evidence of music-making goes back at least 40,000 years ago to the Upper Paleolithic Era when humans were first arriving in Europe.
- Two conflicting ideas about the origins of music point to music either as an "invention," something that piggybacked onto other brain functions, or as biological or evolutionary in origin. The various theories about music as evolutionary suggest a proto-musical language for communication that was important in mother-infant bonding and in group bonding.
- The early proto-musical language eventually split into language to communicate facts and ideas and music to communicate emotion.
- Researchers in the origins of music refer to "music" as cultural product, and to "musicality" as the set of cognitive and biological traits that make it possible to make music. This definition of "musicality" is different from the way in which musicians use the term—to refer to bringing expressiveness to musical performance and to being able to communicate the emotional essence of the work.
- The voice was no doubt the first musical instrument, followed closely by, or perhaps simultaneously with, drumming either on the body or other available surfaces. Instruments followed, as did awareness of the resonant qualities of caves.
- We can't know exactly how early humans played the flute, but it is possible to know from replicas how 40,000-year-old flutes sounded.

3

Born for Music

We grown-up people think that we appreciate music, but if we realized the sense that an infant has brought with it of appreciating sound and rhythm, we would never boast of knowing music. The infant is music itself.

—Sufi master Hazrat Inayat Khan (1882–1927)

Most adults can instinctively find the downbeat in music, the accented first beat of a bar of music, even without musical training. We tap our feet to music, nod our heads, snap our fingers, and sometimes we dance. The ability to detect a beat in music is known as *beat induction*, and that's what makes it possible for us to *entrain*, or synchronize to a beat—to clap together, to dance to music, to play with others in an orchestra, or sing in a choir. We never think about how we are able to find the downbeat, we just do.

The musical abilities of babies

Astonishingly, newborns also recognize the downbeat. Babies move to music as soon as they can move their limbs. It doesn't matter what kind of music, it could be rock, blues, classical, a drumbeat, or hip-hop. Babies are indiscriminate in their tastes. If there is a beat, they move, and it's a whole-body experience for them. They move their arms, legs, torso, and head. They aren't synchronizing to the beat because they don't yet have the muscle control to do so, but they are hearing, and recognizing, the downbeat.

The Musical Brain. Lois Svard, Oxford University Press. © Oxford University Press 2023.
DOI: 10.1093/oso/9780197584170.003.0003

Rhythm

Researchers used to believe that the ability to perceive a beat was acquired during the first year of life by parents rocking their infants to music, but they have now found that two-to-three-day-old babies can detect when the downbeat is missing in a rock rhythm.[1] In a study exploring the perceptual capabilities with which infants are born, babies first heard a simple two-measure, four-beat rock rhythm pattern with snare drum, bass drum, and hi-hat played through couplers on their ears. They then heard variants in which the downbeat, the second, or the fourth beat in a four-beat pattern were missing. The researchers measured brain response with EEG (electro-encephalogram) through electrodes placed on the babies' scalps (Figure 3.1). The babies didn't respond to an omission of the second or fourth pulse, but there was a brain activity response when the babies expected to hear a down-beat but did not.

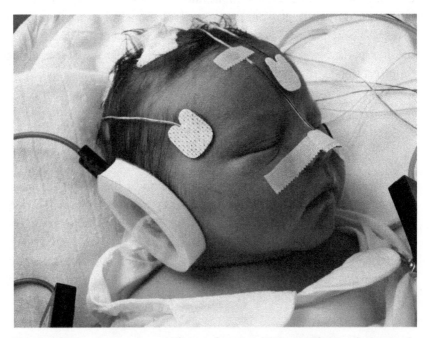

Figure 3.1 Sleeping newborn infant undergoing EEG recording at the hospital ward of the First Department of Obstetrics and Gynecology, Semmelweis University, Budapest, Hungary.
Credit: Gabor Stefanics, PhD.

After hearing the original rhythm, the brains of two-day-old babies already knew when to expect the downbeat and knew when it wasn't there. This is rather astonishing, and it isn't important just in terms of music processing. The auditory abilities that underlie beat induction also allow an infant to adapt to the rhythm of speech and therefore are important in speech processing as well.[2]

This remarkable ability of newborns to recognize when a downbeat is missing suggests that either beat induction is learned in utero or humans are hardwired to detect the musical beat. Given that every baby has a different in utero experience in terms of hearing music, with some being exposed to very little, being hardwired to detect the musical beat seems far more likely. And this is one of the skills that researchers say is necessary for the evolution of music.

Those who study the origins of music agree that there are two musical skills that are fundamental to the evolution of music. They are primary examples of cognitive traits that are biological or hardwired. One skill is beat induction, being able to pick up a beat. The second is being able to recognize a melody no matter what pitch it starts on, referred to by scientists as *relative pitch*.[3] To most adults, these skills seem trivial because nearly everyone has them. But they are not trivial, and *infants have both skills*. Newborns' brains are already prepared to make and enjoy music.

An aside: in the last chapter, it was noted that scientists and musicians do not use the term *musicality* in the same way. The same is true of *relative pitch*. For the researcher, relative pitch is recognizing a melody no matter what pitch it begins on, for example, recognizing "Happy Birthday" in any key. For a musician, relative pitch means that, given a reference note, you can identify or re-create another pitch and identify the interval between them. This contrasts with *absolute pitch*, more commonly called perfect pitch, which is the ability to identify a specific pitch or create a pitch without hearing a reference tone.

Moving to a beat depends on a strong connection between two systems, auditory and motor. The auditory system becomes functional at about twenty-five weeks' gestation and the period from then until about five or six months of age is the most critical to its development. Motor control lags further behind, so a newborn cannot move to music until some control has been gained over motor movement. In a study designed to look at infants' movements to music, 120 infants aged five to twenty-four months listened to excerpts from the last movement of Mozart's *Eine kleine Nachtmusik* in two versions, the original string instrumental version and in a rhythm-only version; to the finale of Saint-Saëns's *Carnival of the Animals* in an instrumental and a rhythm-only version; to a children's song; to drumming in a regular beat; and to a fluctuating rhythm pattern. In addition, half of the infants listened to adult-directed speech, and the other half to infant-directed speech (IDS, also known as "baby talk" or "motherese").[4]

While the children were listening, they were held by a parent who was wearing headphones through which the parent heard spoken speech, so as not to inadvertently influence the child's movements. The infants' rhythmic movements were monitored by observers looking at videotapes of the sessions, but they were also measured using 3D motion-capture technology, which maps movement time onto musical time.

The results confirmed what every parent or caregiver knows—babies love to move to music, but not speech. They moved to the instrumental versions of the musical works, they moved to the rhythmic versions of the works. But they did not move to speech—with one exception. The five-to-seven-month-olds in one part of the experiment moved as much to infant-directed speech as to the musical or rhythmic excerpts. This is perhaps not surprising, since IDS or "baby talk" is more musical, with exaggerated pitch contours (higher and lower), slower rate of speaking, longer vowels, and a larger dynamic range.

Infants between five and twenty-four months were unable to synchronize exactly with the beat because they didn't yet have the muscle control to do so, but the faster the music, the faster they moved their limbs, as though they were trying to synchronize. And the closer they came to synchronizing, the more they smiled. Moving to music makes infants happy, as it does most adults.

Synchronization

It has generally been thought that children are not able to synchronize to a beat until the age of four or five. But with the increasing belief that music originated as a group social activity (see Chapter 2), researchers wondered if children might be able to synchronize at an earlier age if they were in a social situation. A study was designed in the form of a game for children in three age groups: two and a half, three and a half, and four and a half years of age. The name of the study was Drum King.[5]

A researcher played the role of the Big Drum King, and the child was the Small Drum King. The child was first asked to drum along to a radio-like mp3 player. The child was next asked to drum along with a video of a drumming machine. Then the Big Drum King asked Small Drum King to play along with him. Although this may have been a fun game for children, researchers discovered significant information. Children as young as two and a half years of age spontaneously adjusted their drumming tempo to match the beat, but only when they were playing with Big Drum King, not the recording or drum machine. This was earlier than had been thought possible, and all the children, regardless of age, synchronized better when drumming with Big Drum King.

This is not surprising because children learn by imitation. A great deal of visual information, as well as movement information, is conveyed when watching someone perform actions that we want to perform, as we will see in Chapter 9. Some children as young as three are capable of very intricate synchronization—when they learn by observation. Three-year-old Kazuma plays taiko. He didn't learn in isolation. Beginning at the age of two, he watched and imitated a taiko master and played with his father. He learned to synchronize to a beat and drum intricate patterns at such a young age because he was watching and playing with experienced adults. *[You can see a video of Kazuma playing taiko at item 3.1 on the companion website ▶.]*

Embodied cognition

We listen to music with our muscles.

—philosopher Friedrich Nietzsche

Infants may detect the downbeat in their brains, but they feel music in their bodies. When six-to-seven-month-old infants were bounced either on every second or every third beat in a repeating six-beat pattern, they later preferred listening to the auditory version of the pattern to which they had been bounced. The infants who had been bounced on every second beat preferred that auditory version and listened longer to it (one testing method for young infants relies on how long they listen to a particular stimulus). Infants bounced on every third beat listened longer to that version. They preferred the version they had internalized in their bodies.[6]

With the infants blindfolded, the results were the same. They still pre-ferred the version to which they had been bounced. But when they watched a researcher bouncing on every second or third beat without bouncing them-selves, they had no preference. They had to move to the beat themselves in order to internalize the beat, to feel the beat within their bodies. This is *embodied cognition.*

Mind/body dualism, as proposed by French philosopher René Descartes (1596–1650), held that the mind and body, or mental and physical, were en-tirely separate, and that the mind could affect the body, but the body couldn't affect the mind. Dualists today don't believe the mind and body are entirely separate, but they also don't believe the body influences the mind. Embodied cognition, on the other hand, holds that cognition is strongly influenced by the body as well as by the mind. Because human bodies have sensorimotor capabilities that interact with the environment, the body feeds information to the mind both through movement and via the senses, and this feedback influences cognition just as the mind influences the body. Over the past few decades, embodied cognition has become an increasingly important area of research in cognitive psychology, neuroscience, and even the field of robotics.

The study of embodied cognition began to extend to music at the begin-ning of the twenty-first century as a research area within musicology. Music is given meaning through movement—both in perception and performance. We see that in the above study when infants preferred an auditory version of a rhythm to which they had been bounced, preferring what they had physi-cally experienced.

Performers have always known that how we use our bodies affects not only the music we make, but how we think and feel about that music. As Duke Ellington said, "It don't mean a thing if it ain't got that swing." If you can't feel the music, you can't give meaning to the sound. Many music educators have advocated moving to music before learning notation. When I took a class in Baroque music, we were taught to dance the allemande and the courante, typical dance movements found in instrumental suites during the Baroque era. There proved no better way to learn how to perform those dances on an instrument than to feel them in your body—knowing with the body. Frequently, I have had students dance to a rhythm that is giving them trouble at the keyboard. Once they feel it, without fail they can play it.

Someone may say, "I simply cannot move to a beat. I can't feel it," and there are a few people who are beat-deaf, people unable to synchronize their bodies to rhythms they hear. We looked earlier at *amusia,* being "tone-deaf" and unable to distinguish between musical pitches despite musical training. *Beat deafness* is considered a form of amusia, though it is extremely rare. Most people who believe they can't move to a beat simply didn't have enough experience in childhood. Infants instinctively move to music. As they become toddlers, then preschoolers, and then enter elementary school, they should be encouraged to continue to feel music in their bodies and move to it.

Recognition of complex rhythms

People dance, sing, and clap to music in every culture, but various cultures do not value or use the same rhythmic patterns. Music in the Western European tradition, including American popular music, jazz, and rock, is, for the most part, in a simple 2 + 2 rhythm. And though we may not hear them often, almost everyone recognizes that a waltz has a pattern of 3. But music in other parts of the world is organized with different rhythmic patterns. For example, Eastern European music, although sometimes in a simple meter, is frequently in a complex or asymmetrical meter. Five beats in a measure may be arranged in groups of 3 + 2 or 2 + 3. Seven beats in a measure will be found in groups of 2 + 2 + 3 or 2 + 3 + 2. The music of Hungarian composer Béla Bartók often uses asymmetrical meters, as shown in Figure 3.2.

Researchers compared the reactions of babies and adults to variations on simple and complex rhythms to see if the music of one's native culture influenced the ability to process rhythms.[7] Using three groups of subjects— Canadian and American college students, first- or second-generation Bulgarian and Macedonian immigrants, and six- and seven-month-old infants—they played for each group four folk dance tunes from Serbia and Bulgaria. Two were in a simple meter similar to what North Americans are

Figure 3.2 Béla Bartók, Six Dances in Bulgarian Rhythm, No. 6, from *Mikrokosmos*, Vol. 6

accustomed to hearing (2 + 2 + 2 + 2), and two were in a complex meter (2 + 2 + 3), quite common in Eastern European folk music.

The groups then listened to variants of each of the songs. The variants all added an eighth-note beat so that nine eighth notes instead of eight were now in the simple-meter tunes; eight eighth-note beats instead of seven were in the complex meter tunes. The students could identify when the simple meter tunes changed but could not discern when the complex meter was altered. The Eastern European subjects could identify all the changes because they were accustomed to hearing music in both simple and complex meters. And surprisingly, the infants could identify all the changes as well. The infants were able to distinguish—at six months!—all the variants in both the simple and the complex meters, a remarkable skill.

But the real surprise came when the experiment was repeated with different infants who were twelve months old.[8] These babies were no longer able to distinguish the variants in the complex meters. By the time infants were a year old, they had become enculturated to the music of their native country—in this case, the 2 + 2 simple meters heard in most American and Canadian households. However, after a two-week training period of listening to the complex meters every day, the infants could again distinguish the variants in both simple and complex meters. A similar training period didn't seem to help the college students. The old cliché appears to be true: use it or lose it.

Researchers have shown that all newborns have the capacity to learn any language and the ability to distinguish speech sounds in any language. But by the time children are a year old and have been hearing only their native language, the window has narrowed, and they are responsive to a much narrower range of speech distinctions.[9] Pathways in the brain that are not being used have been pruned. That explains why learning a second language becomes more difficult with age. The same appears to be true for music. An ability to distinguish complex meters at the age of six months is lost by twelve months, and even though the infants were able, with training, to again distinguish the variants in the complex meters, college students, with the same amount of training, could not.

Because of what they hear every day, children in other cultures grow up understanding, and feeling in their bodies, the complex rhythms of their native musical traditions. For example, a child growing up in Sub-Saharan Africa will be accustomed to layers of rhythm and three beats in the time of two. The interlocking rhythmic layers of the gamelan will be second nature to an Indonesian child. The kinds of varied-beat rhythmic cycles found in Indian classical music will be familiar to a child raised there. But as learned in the previous cited study, before children become locked into the music of their own cultures, they have the ability to comprehend the rhythms of any culture. *[You can see a video on the companion website of five-year-old Isaiah playing complex Malian rhythms on the djembe, item 3.2. ⊙.]*

Melody and emotion

Infants' response to rhythm in music is physical; they want to move. Their response to melody is emotional. Throughout the world, adults sing soothing melodies to calm or comfort their babies or sing upbeat happy songs for fun and play. Singing is sharing emotions: the emotion of the melody is the message. This is also true for IDS, or infant-directed speech, also used worldwide. Called "motherese" or "parentese" by researchers, IDS is commonly known as "baby talk."

IDS is a style of speech that most people slip into quite naturally when speaking to a baby. Vocal pitch goes up, a much wider pitch range is used (highs to lows) along with a sing-song inflection, vowels are elongated, timbre (tone color) changes, and emotional tone is exaggerated and more highly charged. People speak more slowly to babies, use shorter, simpler sentences, and make up words that resemble the ones they are substituted for, such as "da-da" for daddy, or "wa-wa" for water. Infant-directed speech is used by parents everywhere and most people use it quite naturally. It has been used for generations, and researchers, including Steven Mithen, point to the proto-musical language used by our prehistoric ancestors as a precursor to modern IDS.[10] As previously noted, the proto-musical language, with its variations in pitch, rhythm, and timbre, was an important tool used by mothers to bond with infants prior to the development of language. IDS is still one way caregivers bond with babies and share emotions. Infants prefer IDS to adult speech; they prefer singing to IDS.

Newborns also have musical preferences. Two-day-old babies who were presented with audio recordings of adult and infant-directed speech listened longer to the IDS.[11] What seems to appeal to them are the pitch variations and emotionally charged tone. IDS captures their attention and their emotions. Even when a song is sung to an infant in both an adult way and an infant-directed way, the infant prefers the infant-directed version. They respond to the emotional calming content as well as to the often happy-sounding quality of IDS and play songs.[12]

Singing keeps infants calm longer than speech, even IDS. Researchers at the University of Montreal and University of Toronto compared how long IDS, adult speech, and singing kept an infant engaged. Seven-to-ten-month-old infants from French-speaking households in Montreal listened to recordings of baby talk and adult speech in an unfamiliar language, Turkish, followed by recordings of play songs, also in Turkish. The music or speech continued until the infants showed what is termed "cry face," the universal expression for an unhappy or distressed baby when the baby's face puckers up just before crying begins. When listening to Turkish play songs, the babies in the study remained calm an average of nine minutes before showing "cry face" but only four minutes when hearing Turkish IDS or adult speech.[13]

A different group of infants listened to recorded IDS, adult speech, and play songs in their native French with the same results, staying calm longer to the French play songs than to French IDS or adult speech, although this time they listened for only six minutes to the play songs. The shorter time infants listened to French songs as opposed to Turkish songs may have to do with what has been learned about novelty preference in infants. Very young babies are comforted by hearing something familiar, but as they become older, they prefer something new to the familiar. Two-month-olds prefer a familiar melody; by six or seven months they want to hear new tunes. Turkish was less familiar to them than French, so they listened longer. Infants appear to be sophisticated listeners.

Why do babies prefer singing? For the same reason that they prefer IDS speech over adult speech—because it carries emotion. Infants are particularly attuned to the emotions of singing or IDS speech. They respond to the prosody, the rhythm, stress, and intonation that give meaning, the emotion. For adults, prosody often gives clues about meaning. For example, "Well done" said with a certain emphasis can be genuine praise; with a different tone of voice, it can ridicule.

For infants, the melody is the message, and they particularly like it either happy or calming. Infants as young as five months can pick out a happy song from among a group of sad songs. By nine months, they can do the opposite, pick out a sorrowful song from among a group of happy tunes.[14] Infants are attuned to the emotional content long before the factual or verbal content of the message. This is another argument for the idea that music originated before language.[15]

Musical memory

A good memory for music is another ability in the catalog of infant abilities. Two-month-olds can remember a short melody after hearing it several times and can distinguish it from a similar but unfamiliar melody.[16] Four-month-olds can remember a tune heard in utero. In a Finnish study of twenty-four women conducted during the final trimester of pregnancy, half of the women played "Twinkle, Twinkle Little Star" to their fetuses five days a week for the last few weeks of their pregnancies, half did not. The brains of the babies who had heard "Twinkle" in utero reacted more strongly than the control group to the melody up to four months after birth, as measured by EEG.[17]

Seven-month-olds have a good memory for Mozart. In a study at the University of Wisconsin, Madison, seven-month-old infants listened once a day for two weeks to slow movements from two Mozart piano sonatas (K 281 in B-flat Major and K 282 in E-flat Major). For the next two weeks they heard no Mozart. They were then tested on excerpts from the middle of the movements they had previously heard, as well as excerpts from two other Mozart sonatas they had not heard (K 280 in F Major and K 283 in G Major). They preferred the sonatas they hadn't heard in the first part of the study.[18] They recognized the excerpts from the slow movements they had heard for two weeks, and since they were well acquainted with these excerpts, they preferred the new Mozart excerpts (novelty preference). Not only do babies have a memory for music, they are highly selective listeners.

Sound processing

Any sound that reaches the human ear, whether music or speech, consists of the acoustical elements of pitch, timing, and timbre.[19] Infants are attentive to all three of these elements. They prefer IDS or singing to normal speech because the pitch tends to be higher, so they are aware of pitch differences. Synchronizing to a beat is a matter of timing, and we have seen that they can do that by the age of two and a half. Timbre (or tone color) is that difficult-to-define quality that makes voices sound different, raspy vs. breathy, for example. Mothers shift the timbre of their voices when speaking to infants, using one "voice" when speaking to adults and a different one for infants.[20] Infants respond to the timbre shift.

In the Auditory Development Lab at McMaster University, Director Laurel Trainor and her colleagues had two groups of four-month-old infants listen to CDs of children's songs for twenty minutes a day for a week. One group heard all the songs played on a guitar timbre; the other group heard a marimba timbre. After a week, they recorded the infants' brain activity using EEG while the babies listened to pitches in both timbres—pitches not in the previous week's songs. The babies who had heard guitar timbre displayed larger brain responses to the pitches in guitar timbre, and the babies who had heard the marimba timbre had larger brain responses to the marimba timbre, even though the pitches heard were different from those on the CD. In that short amount of time, not only had the babies learned to recognize the difference in timbre, their brain responses had changed based on the sounds they had been hearing.[21]

Musicians spend years fine-tuning responses to these elements of sound—pitch, timing, and timbre—and become very good at playing "in tune"; being "in time" with another musician, the conductor, or the beat; and recognizing, and producing, fine gradations in timbre vocally or on an instrument. But infants at birth already have some sophisticated responses to these elements.

Other musical abilities

At eight to nine months, infants can recognize a transposed melody, such as "Happy Birthday" begun on any pitch, if the relationship between the pitches remains the same.[22] Infants at five to ten months also recognize a melody when the tempo changes.[23] They can recognize a familiar lullaby or play song in any key and at any tempo. Most adults can sing along to familiar tunes in any key and at different speeds, but these are rather amazing skills for infants.

Early research concluded that infants preferred consonance over dissonance, suggesting an innate preference based on how brains are wired. But just as Western babies can detect variations in asymmetrical rhythms used in Eastern European music, young infants have the capacity to make distinctions and value the music of any culture. Croatian infants find the close seconds used by native folk singers (what Westerners would call dissonance) just as unremarkable and enjoyable as Western babies find the consonant harmonies in the music they hear.

A recent study in Canada confirmed that six-month-olds don't prefer consonance over dissonance. They prefer whichever they have heard most recently.[24] What is true of one-year-olds in terms of preference for familiar native rhythms is true for preference for consonance and dissonance as well. At six months, infants are open to rhythms and harmonies of all cultures. By the time infants are a year old, however, they appear to have become enculturated to the rhythms and harmonies of their own culture.

Language vs. music

Infants have an astonishing range of musical abilities in the areas of rhythm, pitch, melody, timbre, harmony, and memory. What happens between those early months and the age of five or six when many children begin music lessons? Why aren't children as fluent musically by first grade as they are verbally? The short answer is that language has cultural primacy because it transmits concrete facts and ideas. Music, which communicates emotion and feelings in less concrete ways, is relegated to secondary status. As a society we believe it is important to make conversation with children, but we don't attach that same importance to making music with children.

Anthony Brandt and colleagues argue, however, that "without the ability to hear musically, we would be unable to learn language."[25] They maintain that infants' abilities in speech perception are dependent on being able to discriminate the *sounds* of language, and those sounds are musical. An infant's introduction to communication is through a caregiver's IDS and singing, both of which are inherently musical with more pitch, rhythm, and timbre variations than speech itself. It is an "infant's attention to *all* of the musical features of speech [that] provide a richer context for language induction."[26] Music is what makes language possible, and music is what infants hear first.

How children learn language and music

We all grow up hearing language, sometimes for several hours a day. By first grade, children can speak in complete sentences and express complete thoughts. They have a vocabulary of 5,000 words or more. We can understand them well, and they have a reasonable command of language by ear without knowing any rules of syntax and grammar and without knowing how to read or write. By the time they are introduced to letters and words, they already have a template in their brains for the *sound* of language, for phrases, for words, for meaning of sounds—a template on which they can attach the spelling of a particular word. No one would ever consider trying to teach a child to read if he doesn't know how to speak.

Contrast that with music. Infants clearly prefer baby talk (IDS), which is musical. They prefer singing, which has more pitch variance than IDS. Most parents speak baby talk and sing to their infants for the first few months of life. By the time the infant becomes a toddler, adult speech begins to be used and singing all but disappears. A child entering school has heard language and interacted with parents and language since birth. Music was heard in the first few months but then probably not again in any regular interactive manner (just listening to music doesn't count). We often try to teach music to children who are five or six years old by teaching music notation, but they don't have the sound of music in their minds because they haven't been consistently hearing it. Is it any wonder they have difficulty with music?

Annie Jessy Curwen was an Irish piano teacher who taught in Dublin in the late nineteenth and early twentieth centuries. She published multiple music books and essays under the name of Mrs. J. Spencer Curwen, including a Pianoforte Method in 1886 that eventually ran to twenty editions. In an 1898 article for *The Parents' Review*, she wrote that a child's musical education has two stages, one which we think about and one which we do not. Formal teaching begins about age six, and we think about it because we are usually paying for it. But the more important stage is the informal learning that happens from birth to age six, and that stage is likely to be ignored.[27] *[A link to this article is found on the companion website, item 3.3 ⊙.]*

Curwen also comments that we tend to sing to infants to get them to go to sleep, but once they learn to sleep without singing, we quit. And that hasn't changed since 1898. We still sing and use IDS until children are one or two, and then we switch to adult speech, which they continue to hear and use daily. By the time they enter school, having had little to no interaction with music for years, their music abilities are woefully underdeveloped and lag behind their language abilities.

Learning aurally before learning notation

Bach, Beethoven, and Mozart all learned music aurally. They heard music and singing from very young ages and the sounds of music were internalized in their minds. There were no method books. Musical passages were learned by ear, student imitating the teacher, or the student reconstructing music on a keyboard or violin after hearing it multiple times. Later, notation was easily attached to the sound templates in the student's brain. In fact, Mozart often said he heard complete compositions in his mind before ever writing them down.

Prior to the mid-nineteenth century, students practiced scales and arpeggios as a means of learning a musical vocabulary and not as an end in themselves, as is usually done today. As students became more advanced, the teacher would encourage them to invent their own music and to improvise. But the advent of commercially viable printing in the mid-nineteenth century changed that. Method books could be mass-produced, and children were given music books that emphasized the proportionality of note values—whole notes divide into two half notes, half notes into two quarters, etc. Students began learning pitches according to where they were positioned on the staff rather than what they sounded like. They learned fingering according to printed notation, not sound.[28]

Books of exercises, such as Carl Czerny's numerous études for piano, or Rodolphe Kreutzer's studies for violin, became common, and studying a musical instrument changed. Students no longer developed skills in improvisation, composition, and interpretation. They instead began to practice endless hours of technical exercises, and studying music became about reproducing what was notated, about technique and virtuosity.

The tradition of learning music aurally disappeared, even though prominent teachers throughout history have advocated learning by ear rather than from notation. It's ironic that no one has ever suggested that, just because printed texts are available, children should learn to read before they have any sound of language in their ear. Adults would never think of learning a second language from textbooks rather than by listening, yet that happens all the time with music. Learning music by learning notation first is the equivalent of learning a foreign language by studying how words look without ever having heard them. It simply doesn't make any sense. And yet, there remains an ongoing debate about whether notation or understanding by ear should come first in music education.

Sound before sign

Today, listening and singing prior to learning notation is often called "sound before sign" (sometimes "sound before symbol"), and it is sometimes promoted as though it's a new way of learning music when in fact, every musician in the seventeenth and eighteenth centuries first learned music by ear. During the twentieth century, learning "by ear" began to be disparaged and it remains so today as if inferior to learning notation first. The opposite is true. You can't make music if you can't hear it in your mind, and that is the essence of learning by ear. Music educator Edwin Gordon used the term *audiation*, the ability to hear and comprehend, or assign meaning to, sounds that exist in our minds, whether performing from notation, composing, improvising, playing "by ear," or even listening to music. We perceive sound, but audiation is a cognitive process by which the brain gives meaning to musical sounds.[29] Audiation has to do with giving meaning to, or coming to understand, all aspects of musical sound: pitch, rhythm, meter, timbre, volume, style, and tonality. Audiation gives meaning to musical sounds in the same way that we give meaning to the sounds of words in language. Infants have abilities in pitch, rhythm, and timbre from which a sophisticated sound template can be built.

Well-known teachers throughout history, even without having today's understanding of the brain, have known that ear learning must come first. Lowell Mason, America's first public school music teacher (1792–1872), advocated for children to experience music before attempting to learn notation. Austrian composer and educator Émile Jaques-Dalcroze (1865–1950) promoted "good flow" through music and movement. Feeling music through movement, *embodied music cognition*, has been shown to exist in babies when at a very early age they are able to respond rhythmically to music. Adults shouldn't need to be retaught, as Dalcroze Eurhythmics does, to learn to feel music in their bodies.

Zoltán Kodály (1882–1967), Hungarian composer, ethnomusicologist, and pedagogue, created a philosophy of music education that introduces children to musical concepts through listening, singing, or movement, and after becoming familiar with the concept, the child then learns how to notate it. His Kodály Method is not unlike how children learned music in the seventeenth and early eighteenth centuries. The German composer and educator Carl Orff (1895–1982) developed his educational ideas in the "Orff Schulwerk" (Schoolwork), a collection of elementary pieces that combine movement, singing, playing, and improvisation. He treated music like language, believing children can learn without formal instruction through experience and with a sense of play. Composer and writer Anthony Brandt refers to music as "creative play with sound."[30] This is how children should be learning music.

The name Shin'ichi Suzuki is familiar to many for his Suzuki Method, developed after World War II. Children learn to play by ear, notation is added later, and the method relies on parental involvement. Suzuki created his method after struggling to learn German. He noticed that all children achieve some proficiency in their native language prior to entering school, so why shouldn't the same be true for music if they hear it and are involved in it with parents from an early age?

The American Edwin Gordon (1927–2015) was the first music educator to engage in systematic research and field testing to see how children learn music. He concluded that "music is as basic as language to human development and existence."[31] He advocated for singing first, notation second. His *Learning Sequences in Music* is a comprehensive look at how music is learned, what skills are possible at different ages, and what we need to recognize about *how children learn* in order to teach music effectively. Like all the educators profiled here, he emphasized movement and singing in a play-like atmosphere.[32] Consider the loss of human potential when young children do not regularly have interactive exposure to music.

Music lessons for infants?

Since infants have such a range of musical abilities, how do we ensure that those abilities are not only maintained, but expanded and improved between the ages of infancy and when they enter school? Should infants have music lessons? Laurel Trainor and her colleagues at the Auditory Development Lab at McMaster University have studied many aspects of musical development in infants and children. One study was designed to see what effect music classes had on infants' enculturation to Western music. In other words, did music classes for infants have an effect on their sensitivity to the pitch and rhythms in Western traditional music?[33] Six-month-old infants and their parents were randomly assigned to one of two music classes. The "active" class was based on Suzuki Early Childhood Education classes. The infants and parents learned lullabies, nursery rhymes, action songs, movement activities, and played little percussion instruments, as shown in Figure 3.3. Parents participated fully in all the music activities. CDs from the classes allowed them to repeat the activities daily at home.

Infants in the "passive" classes listened to music from the Baby Einstein CDs while they were playing with their parents at play stations with various

Figure 3.3 Infant and dad in participatory music class
Credit: Auditory Development Lab, McMaster University

blocks, balls, and books. Parents used the CDs at home while repeating the activities each day. At the end of six months of classes, infants who had been making music interactively with their parents demonstrated earlier sensitivity to tonal pitch structure. They preferred a version of a Thomas Atwood (1765–1838) sonatina that was played as written versus one that had extra accidentals added and strayed from tonality. They were already becoming enculturated to the tonal organizations and structures of Western music. Brain imaging of the infants, as shown in Figure 3.4, showed earlier or larger responses to musical tones.[34] Their brains had changed because of the interactive music making with parents. One's brain changes as a result of practice, and this can begin even with infants a few months old.

Perhaps in a more surprising discovery, infants in the interactive classes displayed better early communication skills. They smiled more, had lower distress levels in unfamiliar situations, and were easier to soothe when upset than the infants in the passive classes. Though infants in both the active

Figure 3.4 Infant's brain responses to musical tones being measured with EEG
Credit: Auditory Development Lab, McMaster University

and passive classes listened to music, listening wasn't the important thing, it was the interactive participation with a parent, making music together as infants do when they interact with parents with language. The quality of the singing or playing doesn't matter. Just as infants don't have the muscle control to synchronize, they also don't have the muscle control to play an instrument accurately or to produce pitch accurately. We don't expect a two-year-old to sound like an adult when using language. Neither, at that age, will their music-making sound very much like what we consider music. But they will be absorbing sounds and creating a "sound to meaning" connection for music in their minds. They will be learning to audiate. When I was a child, an aunt spent a good deal of time at our house, always at the piano, playing and teaching us songs. Our babysitter also taught us a lot of fun, nonsense songs, many of which I remember to this day. When I began piano lessons at age five, learning notation was never an issue because it was easy to attach the notated symbols to the sounds I already had in my mind. I had already learned to audiate at a simple level. The ability to audiate more complex musical sounds and structures increases as one's skill level increases.

Sensitive period for learning music

Sarah Watts is a specialist in early childhood music education at Penn State University. She teaches her graduate students how to work musically with infants and toddlers. The infants and toddlers are eager participants, toddling over to the guitar and trying to help Dr. Watts play as she sits on the floor, attempting to shake rattles in time to the music, and even trying to find the coordination to throw scarves into the air in time with the music.

Researchers have generally said there is no sensitive period (what used to be called critical period) for learning music as there is for learning language (to age five), but studies such as the one above from the McMaster Lab indicate otherwise, as do the eager toddlers in Watts's music classes. Musicians are vividly aware that early exposure to music makes a difference. Pianist-conductor Daniel Barenboim speaks about growing up in a household with two parents who taught piano. Everyone he met played the piano or another musical instrument and everyone who came to the house made music.[35] Playing the piano was a natural part of his life as a child and a natural form of human expression. Music was simply in the air. He grew up hearing music, began piano lessons with his mother when he was five, and by the time he was eleven, was playing with the Berlin Philharmonic.

Pianist Ann Schein, who has had a concert life spanning more than sixty years and who was on the faculty of the Peabody Institute for twenty years, says that she was a toddler during World War II. The family lived in Evanston, Illinois, and patriotic music was played on the radio twenty-four hours a day. At the age of three, she played the Marine Hymn—with both hands—at the piano. After hearing it repeatedly on the radio, she could hear it in her mind, and she found a way to pick it out at the piano. Her parents decided that if she could play that well by ear, she should have a teacher.[36] She made her debut in Mexico City at the age of seventeen, performing both the Rachmaninoff Third Concerto and the Tchaikovsky B-flat Concerto.

Learning music is different from learning to play an instrument. Very young children lack the motor control necessary to play an instrument well, although YouTube videos provide evidence of younger and younger children performing on everything from the piano to a drum kit. Nor do they have the motor control to be able to control pitch, so they usually can't sing in tune. But that doesn't mean they can't have fun with music. Children before the ages of five or six can learn to sing songs (to the best of their ability), move to rhythms, play little percussion instruments, and enjoy music as play. Some early childhood teachers put a large staff and cut-out notes on the floor. The preschoolers hear and sing simple songs while notating them on the floor, and then they move along the appropriate lines and spaces as they sing. They are feeling the movement, connecting movement with sound, and sound with notation. Learning music becomes a whole-body experience—and it's fun as well.

Children need to have a template for music, for pitch and rhythm, in their brains as they have for the sounds of language before they can understand and attach notation. Not every child can become a professional musician, but learning to enjoy and participate in music before the age of five will mean more people are comfortable with the language of music, and with being able to express themselves in that language.

You don't need a particular method to use with children. Infants are born wanting to move to music. Keep that love of moving to music alive. Infants are also born responding to singing more readily than to speech. Capitalize on that by singing to and with children so they learn to internalize melodies in their minds. Just as we interact continually with infants and young children with language, we need to do the same with music so they can develop the musical abilities with which they are born.

Key Concepts

- Infants are born with remarkable abilities in rhythm, pitch, melody, musical memory, and emotional response to music, suggesting biological foundations for music.
- Any sound that reaches the ear, whether music or speech, consists of the acoustical elements of pitch, timing, and timbre. Infants have rather sophisticated responses to all three.
- Being able to hear musically is what makes learning language possible. An infant is introduced to communication through a parent's or caregiver's singing and infant-directed speech, both of which are musical, with variations in pitch, rhythm, and timbre.

- Children have a template in their minds for the sound of language before they begin to learn the alphabet or words. Similarly, they need a template in their minds for the sound of music before they can begin to learn notation.
- There is a sensitive period for learning music as there is for language—the first five to six years of a child's life.
- Singing, playing rhythm games, and moving to music with children provide a foundation for later music learning.

4

Learn an Instrument—Change Your Brain

> Your brain—every brain—is a work in progress. It is "plastic." From the day we're born to the day we die, it continuously revises and remodels, improving or slowly declining, as a function of how we use it.
>
> —neuroscientist Michael Merzenich[*]

We hear a lot about skilled athletic performance, about the demands on the body of being an athlete at a professional level, about peak conditioning. Music performance doesn't receive the same kind of attention; it is assumed to be not as demanding. Yet musicians are expected to perform on instruments, the voice included, with power, speed, virtuosity, endurance, and coordination over much longer periods of time than most athletes. In the Olympics, the long program for a figure skater is four minutes, the floor exercise routine for a gymnast is ninety seconds. Baseball players sit on the bench for significant periods of time when they aren't on the field, so their actual playing time is far shorter than the average length of a three-hour baseball game.

In contrast, a violinist playing in the orchestra for Wagner's opera *Die Meistersinger von Nürnberg* (The Mastersingers of Nuremberg) plays almost continually for four and a half hours—over five if the opera is done in an uncut version—usually with two intermissions; a baritone singing the role of Hans Sachs in the opera will be onstage singing for much of the opera. A drummer in a club plays continually for at least two or more hours with only short breaks. A pianist or violinist playing a solo concert performs for two or more hours with one intermission.

In addition to having to be in peak physical condition and capable of impressive athleticism over a longer period of time than most athletes, musicians' athleticism *must be connected to sound*, and the resulting *sound must convey emotion*. Neuroscientists often say that making music is the most complex cognitive activity in which a human engages. Before exploring

The Musical Brain. Lois Svard, Oxford University Press. © Oxford University Press 2023.
DOI: 10.1093/oso/9780197584170.003.0004

that complexity in more depth and seeing how the brain changes as we learn an instrument, let's look at some basic information about the brain.

Brain 101—an overview

Everyone's brain looks basically the same. Each has a central nervous system consisting of the brain and spinal cord. The mushroom-shaped mass of nerve tissue that is called the brain weighs about three pounds and serves as the control center for the entire body. Figure 4.1 shows the three main parts of the brain: the *cerebrum*, *cerebellum*, and *brainstem*.

From an evolutionary standpoint, the brainstem is the oldest part of the brain. Located at the base of the brain, it connects the cerebrum to the spinal cord and cerebellum. All information relayed from the body to the cerebrum or vice versa must go through the brainstem. Because it is the oldest part of the brain, it controls many of the automatic functions present at birth, including breathing, heart rate, blood pressure, body temperature, and swallowing. But it also plays a significant role in how music and speech are processed (see Chapter 10). The cerebellum in the back of the brain coordinates and fine-tunes muscle movements and helps maintain posture and

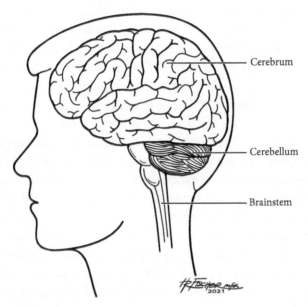

Figure 4.1 Three parts of the brain

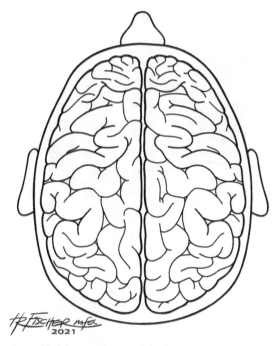

Figure 4.2 Right and left hemispheres of the brain

balance. It is important for motor control and the motor skill learning necessary in playing an instrument.

The cerebrum forms the major part of the brain and is divided into left and right hemispheres, seen in Figure 4.2. They are linked by a band of nerve fibers called the *corpus callosum*, which carries information between the two hemispheres. The right hemisphere receives sensory input and controls movement on the left side of the body, and the left hemisphere receives sensory input and controls movement on the right side of the body.

The cerebrum is covered by a thin, wrinkly layer called the cerebral cortex, or neocortex, where most of the brain's neurons are located. Each hemisphere in the cerebrum is divided into four lobes: frontal, temporal, parietal, and occipital, seen in Figure 4.3. The *frontal lobes* are directly behind the forehead and are the largest of the four major lobes. They are considered the control center for behavior and emotions as well as the home of personality. Frontal lobes are important for language and for controlling higher-level cognitive skills such as working memory, judgment, planning, problem solving, decision making, attention, and impulse control—known collectively as

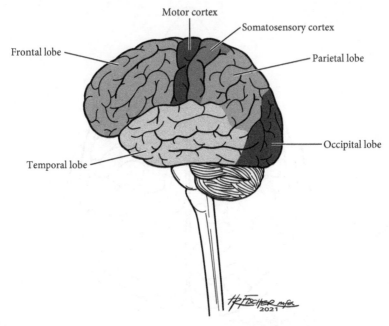

Figure 4.3 Four lobes of the brain

executive functions. The *motor cortices*, at the rear of the frontal lobes, control voluntary muscle movement, the movements with which an instrument is played.

The *parietal lobes* are located toward the back and top of the head. They contain the *somatosensory cortices*, which process sensory information, such as touch, temperature, and pain. The parietal lobes also have a role in processing information about how our body moves as we play an instrument or sing. The *occipital lobes* at the back of the head are primarily responsible for vision. The *temporal lobes*, basically behind our ears, process auditory information. They are also involved with encoding memory and have a role in processing emotions, language, and some aspects of visual perception.

Neuroplasticity

Although everyone has the same brain areas described above, the brains of a musician and an athlete, for example, will display differences. We are what we do—what we study, learn, or experience. Brains develop in a particular way in response to how they are used every day. The specific brain areas that are developed and strengthened in acquiring, practicing, and maintaining the skills of a professional musician are quite different from those of a professional athlete, or for that matter, a simultaneous interpreter or neurosurgeon. Days spent learning different things and engaged in different experiences and activities that require processing by different brain areas change individual brains in significant ways. The brain constantly rewires itself in response to the amount and kind of sensory stimulation reaching it. This ongoing rewiring of the brain is called *neuroplasticity*.

The term *plasticity* itself was first proposed in the 1880s by the American philosopher and psychologist William James. He argued that neuronal pathways in the brain become deeper, wider, and stronger the more they are used, just as pathways in a country road become deeper and more permanent the more they are traveled. In a short treatise titled *Habit*, written in 1887 and reprinted as a chapter in his highly influential 1890 textbook called *The Principles of Psychology*, James explained that if we repeat an action or behavior enough times, it tends to perpetuate itself and will eventually be done automatically. It becomes a habit because the brain changes.[1]

Research by other scientists over the next few years appeared to confirm the idea of plasticity. But Santiago Ramón y Cajal, who initially supported the idea of neuroplasticity, changed his mind, and the view he expressed in his 1913 textbook *Degeneration and Regeneration of the Nervous System*—that nerve paths were fixed and unchangeable in adults—became accepted dogma.[2] Given Cajal's esteemed reputation, his view became widely accepted and remained established doctrine for the next several decades.

Scientists came to believe that, although the young brain was highly plastic, the adult brain was not. Neurons could die, but would not be replaced by new ones, nor could they reorganize, strengthen, or create new pathways in different ways. How, then, to explain the fact that adults were able to learn a new language or learn a musical instrument? Scientists agreed that new learning could happen in adult brains, but assumed it happened using existing brain connections. They were convinced the brain could not change functionally or structurally.

When I was a young piano student, the consensus among my fellow students was that you must develop all the technique you will ever need by your late teens—after that, it's too late. Music pedagogy basically mirrored what the scientific world believed. The brain is hardwired by the time you were in your late teens, an idea that, unfortunately, still surfaces today. That belief didn't make sense, considering our plans to attend graduate school to become better performers, with graduate school extending into our mid-twenties or beyond. It also wasn't consistent with reviews of well-known older performers that noted the refinement of technique and increasing depth of musicianship from early years to later performances. Obviously, these artists were changing for the better over time, but there was no basis for drawing a relationship between brain activity and playing a musical instrument. As far as we were concerned, technique was all about muscles. George Kochevitsky's book (see Chapter 1) talked about practicing with the brain, but it was the only pedagogical text to mention the brain and it was not well known.

The idea that the brain is unchanging in adults also had serious negative ramifications for patients suffering strokes or traumatic brain injuries because very little was done for them in the way of therapy. The medical community recognized that some patients recovered from strokes, but it was assumed recovery was relative to the seriousness of the stroke. In general, rehabilitation programs were short, and patients were sent home or for institutional care because there was no expectation of recovery (more in Chapter 7).

Maps in the brain

A Canadian neuroscientist named Wilder Penfield conducted some ingenious work in the 1930s that eventually served as a foundation for many later studies of neuroplasticity. Penfield, the founder and first director of the Montreal Neurological Institute, spent his career doing groundbreaking research in several areas including hallucinations and illusions. During the 1930s and 1940s, he was particularly interested in epilepsy. At that time, there was no medication available that was effective in preventing seizures; the only remedy was surgery. Penfield conducted hundreds of surgeries on epileptic patients, excising the scar tissue (often the result of a brain injury) where the seizures originated.

To avoid cutting into either the somatosensory cortex or the motor cortex, in which case patients would lose their sense of touch or motor function, Penfield would first stimulate the brain with electrodes while the patient was fully conscious to discover what areas he should avoid. A local anesthetic had been used while a portion of the patient's skull was removed. The brain doesn't have pain receptors, and a patient needed to be awake to tell Penfield what he experienced while Penfield probed his brain.

Over the course of hundreds of surgeries during the next fifteen to twenty years, as Penfield probed brains, patients would report feeling a tingling on the right hand or left knee or lips or elsewhere on the body. Perhaps a patient's knee would jerk, or a finger would move. Penfield would mark the spot in the brain with a small, numbered note and an assistant would record the results. The procedure was not unlike a recent Verizon commercial, "Can you hear me now? Can you hear me now?," as the cellphone user moves to different areas. In Penfield's operations, it was "Where do you feel this now?" "Where do you feel this now?" as he probed different areas of the brain.

Over time, Penfield discovered that the somatosensory cortex and the motor cortex each has a map that is a neural representation of all our body parts. He named this map the "homunculus," a mid-seventeenth century Latin diminutive for "man." He published his findings in a 1950 book, *The Cerebral Cortex of Man*.[3] Illustrations represented the two maps, sensory and motor, and those originals have been redrawn and used many times since then.

As the illustration in Figure 4.4 shows, not all parts of the body occupy equal amounts of brain space. In the somatosensory cortex, the amount of dedicated brain area is relative to where we need the greatest touch sensitivity in the body. Fingers, which are primary touch receptors, occupy far more brain area than toes, which don't have such an important sensory function. Each finger has a separate brain area. Sensory sensitivity is also extremely important in the tongue and lips, which occupy relatively larger brain areas than, for example, the nose.

In the motor cortex, the distribution of brain area is relative to which areas of the body are used in the most highly coordinated ways, such as in making music or sculpting a work of art. Again, hands and fingers have more brain area than knees or ankles. Individual fingers have separate brain areas because they function independently. Interesting new research shows that people who paint with their toes have toe maps in the brain similar to those of the fingers in people who have full use of their hands, a result of how neuroplasticity works.[4]

Initially, Penfield's maps seemed to confirm the belief that adult brains were fixed and unable to change because very precise cortical areas were assigned to specific anatomic structures. But by the 1970s and 1980s, Penfield's maps were being used to help prove the opposite—that the adult brain could

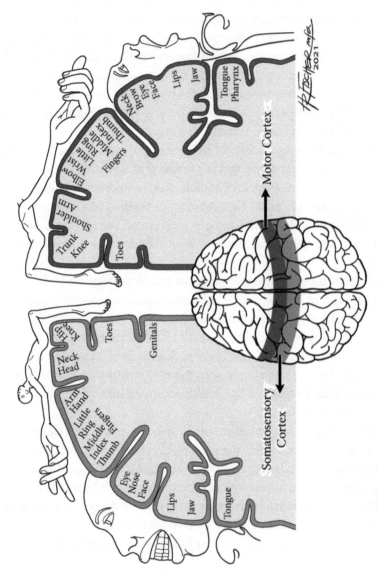

Figure 4.4 Somatosensory and motor homunculi

change in response to learning or experience. In musicians, many of these brain areas change as a result of practice.

Neuroplasticity—not just for kids

Neuroscientist Michael Merzenich believed, as had William James back in the 1880s, that the adult brain could change, that plasticity did not just occur in children. Over a period of twenty years in the 1970s and 1980s, Merzenich conducted a series of experiments with owl monkeys that proved conclusively that neuroplasticity could occur in adults of the species.[5] The next step was to prove that adult human brains could change.

In 1993, a young scientist named Alvaro Pascual-Leone showed that the right forefinger of people who read Braille was represented by a larger brain area in the somatosensory cortex than either the left forefinger or the right forefinger in a control group. This was a sure sign that more sensory input from the right finger resulted in an expansion of the brain area that represented that finger—neuroplasticity in the brains of adult humans.[6]

To measure the brain areas, Pascual-Leone used a technology called transcranial magnetic stimulation, or TMS. TMS is a noninvasive procedure using magnetic fields to stimulate nerve cells in the brain. It has been used to diagnose and treat clinical conditions including migraine headaches and depression. It is also used for the pre-surgical mapping of motor functions. Pascual-Leone essentially used it as a more sophisticated version of what Penfield was doing in the 1930s and 1940s.

Pascual-Leone was a strong believer in the neuroplasticity of the adult brain. He was also a sports fan who played both soccer and tennis. He experienced the improvement of his own skills the more he played, so he decided to explore how the motor areas of the brain changed over time with physical practice. But he didn't use athletic practice; he used practice on a musical instrument. He devised an experiment using right-handed adults who had never played a musical instrument and who had never typed with all fingers. Using a Yamaha electronic piano keyboard interfaced with a Mac IIci computer (this was 1994), each participant practiced a five-finger exercise: C D E F G F E D C using fingers 1 2 3 4 5 4 3 2 1. The participants learned to play this exercise as fluently as possible at a metronome marking of sixty beats per minute, the beats corresponding to fingers 1 and 5 (thumb and little finger).[7]

Before the first day of practice, a baseline measurement was taken with TMS of the area in the left motor cortex corresponding with the muscles controlling the fingers in the right hand, the long finger flexor and extensor muscles (the left hemisphere controls the right side of the body). Because of Penfield's study, that area in the motor cortex was known. Then the subjects in the study practiced for two hours a day for five days. Anyone who has played an instrument knows how tedious this would have been. The subjects were told to try to play fluently, without pauses, and to play the notes as evenly as possible. They were tested at the end of each two-hour session to see if they could do twenty repetitions of the five-finger exercise without errors. Twenty to thirty minutes after each practice session, the subjects again had TMS mapping of the finger areas.

As one would expect, playing skill improved over the five days. The subjects were able to play more evenly and fluently, with fewer mistakes. That's what practice is supposed to do. But in a significant confirmation of neuroplasticity in adult humans, the cortical map corresponding to the fingers of the right hand steadily increased in size each day. Practicing the piano literally changed the size of the motor cortex corresponding to the fingers of the right hand. Pascual-Leone suggests there are two ways this reorganization of the motor cortex can happen. One is the establishment of new connections, and the second is the unmasking of previously existing connections. Since the motor maps changed in such a short period of time, he suggested it was probably the latter, that it was due to the unmasking of connections already in existence. The initial changes were short-lived, and the subjects' brains returned to baseline after a weekend. But this study demonstrated a basis for longer-term structural changes in the motor cortex as the skill becomes more automatic.[8]

There were two control groups. Group 1 did not practice at all, and there was no change in the finger maps of those individuals. Group 2 practiced for two hours each day, but those individuals were told to play randomly, one finger at a time and play anything they wanted. There was a slight increase in the finger maps of subjects in Group 2, but nothing significant. Practicing the piano, or any musical instrument, changes your brain. An aspect of this study that is extremely important to musicians is that intentional practice of a specific pattern is what led to changes in the brain's finger maps. Randomly practicing any notes, although it changed the maps slightly, didn't have the same significant effect. Intentional practice is important for the development of specific skills, as we will explore further in Chapter 6.

Brain areas involved in making music

The motor cortex that figured prominently in Pascual-Leone's study is just one of many brain areas that are necessary for making music. As you read the following sections, you may be overwhelmed by the large number of brain areas involved, but on the other hand, it is fascinating to think about how much of the human brain has evolved to support music. What is described in the following pages is actually a rather pared-down account. The reality is more complex, as neuroscientists will tell you, and researchers continue to discover more intricate ways in which music-making is supported by our brains. Certain areas will be explored in a bit more detail as they relate to a particular aspect of making music, for example, in the discussion of learning and memory in Chapter 5, imagery in Chapter 8, and the auditory system in Chapter 10.

The auditory-motor connection

The connection between sound and movement may be the oldest and most important connection supporting music-making in the brain. The brainstem, in addition to regulating functions including breathing, heartbeat, body temperature, and equilibrium, also controls reflexive actions, such as jumping when startled by a sudden loud sound. Even babies reflexively turn toward loud sounds because that reaction to sound is hardwired in the brain. Some researchers say that, from an evolutionary perspective, the connection between sound and movement developed to keep our primitive ancestors from danger: hear an unexpected noise and respond quickly by moving in the opposite direction to avoid becoming a lion's dinner or being crushed by a falling rock. Early humans who paused to think about it didn't survive. Those who reacted instantaneously were alive to pass on their genes to the next generation, so the link between sound and movement was wired into the human brain a long time ago.

The processing of sound goes well beyond reflex reactions, however. We voluntarily link sound to movement in far more sophisticated and sometimes spontaneous ways: dancing to music, clapping to a beat, chanting in rhythm at athletic events, singing "Happy Birthday," detecting the beat in music and intuitively moving to it. Our prehistoric ancestors were using that sound-movement connection when they were making music 40,000 years ago. For musicians, that sound-movement connection becomes especially strong because of the years spent refining and strengthening it through continual practice.

Figure 4.5 shows the auditory-motor interaction as one plays an instrument.[9] Play a note on a violin or strike a drumhead with a stick and the string or drumhead vibrates, causing the molecules in the air around it to vibrate. The resulting wave of sound enters the ear causing the eardrum to vibrate. The eardrum transmits the sound vibrations to the cochlea, a snail-shaped structure in the inner ear that turns the mechanical energy of the wave into electrical impulses. The auditory nerve transmits these electrical impulses

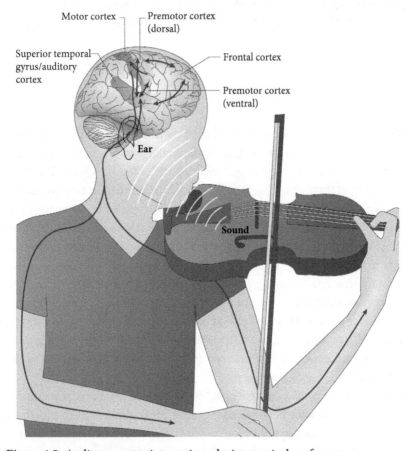

Figure 4.5 Auditory-motor interactions during musical performance

Credit: Figure reprinted by permission from: Springer Nature, *Nature Reviews Neuroscience* 8, "When the Brain Plays Music: auditory-motor interactions in music perception and production," R. J. Zatorre et al.© Nature Publishing Group 2007.

via the brainstem to the *auditory cortex* of the brain. At various points along the auditory pathway, the signal is decoded for duration, frequency and intensity. When it reaches the auditory cortex, the signal is perceived as one of thousands of sounds to which we have learned to attach meaning—music, speech, or perhaps the doorbell (explored in greater detail in Chapter 10).

After processing in the auditory cortex, neural signals are sent to the *supplementary* and *premotor cortices*, responsible for the planning, preparation, and guidance of movement in response to the sound we just heard—whether and how to move the bow arm or strike the drumhead. These cortices then send electrical signals to the *primary motor cortex*, which converts those plans for movement into signals that are sent along neural pathways down the spinal cord and to the muscles in our arms and hands to initiate the movements to create more sound—and the sound-movement loop continues. The muscles can't act on their own until the motor cortex in the brain initiates motion. Practicing happens in the brain, not the muscles. The auditory-motor loop is nearly instantaneous, and it allows us to make music. If it were necessary to think about connecting each sound to a movement as we do when first learning an instrument, performances of great music would never happen.

When learning a musical instrument, sound and the movement necessary to create it become represented jointly in the brain.[10] Researchers have found that when pianists listened to piano music that they themselves had played, there was increased activity in the motor cortex even though they were just listening and not playing. These findings support the idea that there is a close connection between perception (listening) and execution (playing) in the brain.[11] Many musicians experience this when they find their fingers moving while listening to someone else play a piece they have also played.

Conversely, the auditory cortex can be active when you are playing silently. Amateur and professional violinists were scanned using fMRI as they silently fingered the first sixteen bars of the Mozart Violin Concerto in G Major (KV216). The brains of all the violinists showed activity in the motor cortex because they were fingering the music. But the professional violinists additionally displayed significant activity in the auditory cortex because the auditory-motor connection had been strengthened through years of practice.[12]

This connection between auditory and motor areas of the brain is established quickly. In a 2003 study in which adult non-musicians were taught to play a melody on a keyboard with the right hand, activity between the auditory and sensorimotor areas of the brain showed up within the first few minutes. After five weeks of practice, that connection was firmly established.[13] That does not necessarily mean the piece of music was learned in five weeks. It simply means that those two brain areas were already working together. The more practice, the stronger that auditory-motor connection becomes. It forms the basis for the more complex auditory-motor skills professional musicians develop over a period of years. This is neuroplasticity, the brain changing in response to learning.

Ohad (Udi) Bar-David, cellist with the Philadelphia Orchestra, is the founder of Intercultural Journeys, a nonprofit organization that works to provide opportunities for musical dialogue among musicians of various cultural traditions. Bar-David often plays music outside the Western classical tradition in which he was trained. He says that when he first began playing with Arab musician Simon Shaheen, it was difficult to play the microtones that are prevalent in Arab music. "But," he says, "when you start hearing it, your fingers just take you there."[14] If you hear the sound in your mind, the connection between brain and fingers makes it happen.

Other brain areas for making music

The auditory-motor connection may be the oldest brain connection for music-making, but it is far from the only one. Just as the ear is structured to turn incoming sound waves into electrical impulses, the eye is structured to turn incoming light waves into electrical signals that are carried by the optic nerve to the *visual cortex* in the brain where the information is decoded, whether that information has to do with notation, your instrument, the conductor, or musician colleagues. The brain areas discussed in this section can all be seen in Figure 4.6.

Touch receptors in the skin of the fingers and hands gather information and turn it into electrical impulses. Those electrical signals are sent along nerves to the spinal cord and then up the spinal cord to the *somatosensory cortex* in the brain. Different pressures or touches that result in different qualities of sound send different signals to the brain. That information must be processed in sensory areas of the brain and then sent to *premotor areas* that assist in integrating sensory and motor information, and *supplementary motor areas* that plan complex movements. The neural signals then go to the *primary motor cortex* that initiates the movements necessary to produce

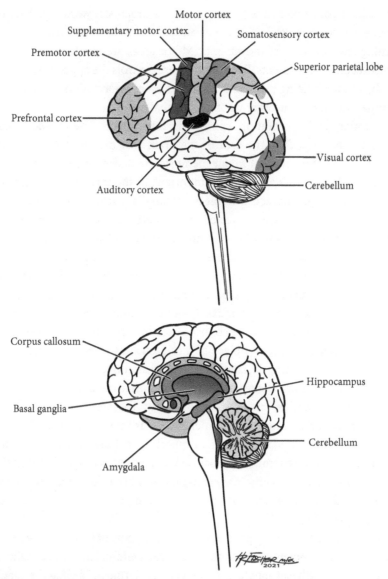

Figure 4.6 Brain areas involved in processing music

those particular kinds of sounds. *All sensory information enters the brain as electrical patterns.* We'll see why this is so important in Chapter 7.

Most instruments require two hands, doing different things simultaneously, so bimanual coordination is important. Many brain areas are involved in making your two hands (plus two feet for organists) work together. The *cerebellum* coordinates and regulates muscular activity and is involved in the timing and accuracy of movements. Motor areas are involved in planning and executing movements. Each hemisphere controls the opposite side of the body, so coordinating two hands requires a lot of information to travel from one hemisphere to the other via the corpus callosum.

Emotion is important in communicating music. The limbic system includes the *amygdala*, the area of the brain that processes emotion and attaches emotional content to our memories, including memory for music. It also includes the *hippocampus*, important in the formation of memory. The limbic system interacts with the *basal ganglia*, a group of structures deep within the brain that are connected to the motor and sensory cortices. The basal ganglia are involved in the automation of skilled movements and in storing fast and automated movement programs, playing a key role in the formation of procedural memory, the memory for playing an instrument, as we will see in the next chapter. The basal ganglia are also involved in the reward system, and we make music because we find it rewarding, or pleasurable.[15] The *prefrontal cortex* controls executive function skills needed in learning an instrument or a piece of music, such as working memory, planning, problem solving, decision making, and attention.

Then there are the elements of music itself that the brain must process, such as pitch, rhythm, melody, and harmony. These are processed in different areas of the brain: basic elements such as intervals and rhythms, for example, are processed in the left hemisphere; more holistic elements such as meter and melodic contour are processed in the right hemisphere.[16] Rhythm and timing are particularly interesting, not just because of their impact on making music, but because they also have an impact on speech processing (see Chapter 10).

Whether we are actively producing a rhythmic beat or just listening to one, premotor and supplementary motor areas, cerebellum, and basal ganglia are all involved. The basal ganglia may, in fact, play a role in "feeling the beat."[17] We talk about being "moved" by music, and perhaps that is because so many brain areas involved in movement are necessary for processing rhythm.

Timing, being able to be "in sync" with someone or with the beat, is processed in many of the same brain areas as rhythm: pre-motor and supplementary motor areas, cerebellum, and basal ganglia. But two other very important structures are the parietal cortex, which has a crucial role in estimating duration of time, and the prefrontal cortex, involved in the perception of time.[18] Musicians use "timing" to refer to keeping a beat and synchronizing with another performer or ensemble. We also use it to refer to "expressive timing," or the liberties we take with the placement or duration of note to achieve expressive effects, such as those frequently referred to as "rubato."

Reading notation

Finally, what about reading music notation? Certainly not all musicians use scores. Jazz players use charts (sketches or notes) or they improvise outright, and in many cultures, music is passed down from generation to generation by ear and no musical notation exists. In Western classical music, scores use symbols to convey the music. Scores contain information about pitch, rhythm, duration of notes, dynamics, articulation, tempo, and more. Reading musical notation is far more complicated than reading text. With some exceptions (including Arabic, Hebrew, Chinese, and other Asian languages), reading text moves horizontally from left to right. Musical notation is not only horizontal for a simple melody, but also simultaneously vertical when reading chords, accompaniments, or full orchestral scores including lines of notation for each instrument. Text has no rhythm or tempo instructions; the reader sets the speed. Music notation specifies rhythm and tempo. Reading music notation involves decoding this spatial arrangement of pitch and rhythms over time. Despite the complexity of reading musical notation, relatively little research has been done to determine how it happens.

When reading a musical score, the brain must translate visual information in the form of symbols into a motor program that specifies patterns for pitches, timing, and positioning of movements.[19] Translating spatial information from a score into complicated motor commands has been found to happen in a part of the brain called the *superior parietal cortex*, which as we just saw, plays a crucial role in timing.[20] Not surprisingly, that area of the brain is also critical for working memory, which allows one to retain the spatial information from a score long enough for the brain to create the motor commands necessary to play the notes and rhythms.[21] Areas important for pitch, rhythm, and the reading of symbols have been found in the occipital, parietal, and temporal lobes, seen in Figure 4.3.[22]

Connection of neural networks

For all the brain areas just discussed to work together to make music, they must be connected. The job of brain neurons is to connect these areas and to process and transmit information throughout the brain, spinal cord, and peripheral nerves to the body. The brain is made up of 86 billion nerve cells, or neurons, and about an equal number of glial cells.[23] Neurons transmit information, while the job of the glial cells is to provide nourishment, support, and protection for neurons. Both are essential.

Each neuron consists of a cell body, multiple dendrites, and a single axon, seen in Figure 4.7. Dendrites bring information to the cell body in the form

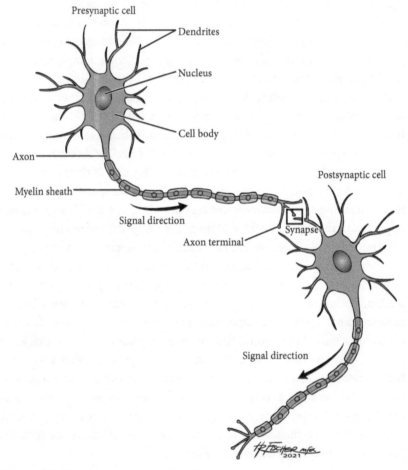

Figure 4.7 Two neurons connecting at a synapse

of electrical signals. The cell body contains genetic information that directs the activity of the cell. A single axon extends away from the cell body and its job is to carry information to other neurons. An axon has multiple endings or terminals. When electrical impulses reach these terminals, chemical neurotransmitters are released across a gap called the synapse to transmit the message to a dendrite of the next cell in the pathway. This electrochemical process continues from cell to cell until a neural pathway is formed extending from one area of the brain to another. Larger networks are formed connecting multiple areas of the brain, such as sensory, auditory, and motor areas. Transmission among all these areas occurs within microseconds.

The 86 billion neurons in the brain can combine in an infinite variety of ways and make networks connecting all the parts of the brain that control thought, behavior, action—and music. See an example of neural networks in Figure 4.8. Some of those connections are formed during the development of the fetus, such as those for life functions like breathing, the beating of our hearts, or the movements of a fetus in the womb. We have seen that some neural circuits for music are already present at birth, such as those for complex rhythms or pitch structures. Other neural pathways are created through experience, learning, or injury—for example, riding a bike, learning a foreign language, learning to play a musical instrument, or recovering from a traumatic brain injury.

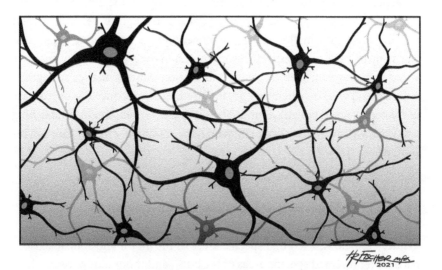

Figure 4.8 Neural networks

How we learn

In 1949, Canadian psychologist Donald Hebb introduced the theory that the more often one neuron fires consecutively to the next along a particular pathway, the more the synapse between them is strengthened, and the more likely it is that the neurons will continue to fire consecutively. This is what happens when we practice and our skill improves. This is synaptic plasticity, or plasticity at the level of a single cell. There are three synaptic processes that enable learning: (1) the synapses strengthen or weaken over time in response to increases or decreases in use—how much or how little we practice; (2) more neurons are recruited as we add more information during our practice, resulting in an increased number of synapses connecting neurons; and (3) a substance called myelin is wrapped around the axon as it is used. Myelin could be compared to the rubber or plastic covering on electrical wire. It provides insulation for the axon and speeds transmission. The more often a signal is carried by the axon, the more myelin is wrapped around it and the faster the speed of transmission. All these synaptic processes are important in the process of learning and memory.

The theory Hebb introduced, called Hebb's Law, Hebbian theory, or Hebbian learning, is usually stated as "neurons that fire together, wire together."[24] Practice facilitates repeated firing of the synapses and wiring together of neurons. When learning a musical instrument, people often say, "Practice makes perfect" but the repeated firing of the synapses *ensures permanence, not perfection*. The brain doesn't distinguish between correct and incorrect information during practice, there is no flashing signal that warns of incorrect rhythms or wrong notes. Wrong information becomes wired just as readily as the correct information. Since synapses strengthen with use and weaken when they are not used, mistakes, once discovered, are corrected by not repeating them so the synapses weaken, and by establishing a new pathway with the correct information, making it stronger through repeated use.

Musicians are models of neuroplasticity

Wilder Penfield showed that specific areas in the motor and somatosensory cortices represented specific areas of the body, and thousands of researchers have shown that various brain areas show neuroplastic changes depending on the activity involved. The cerebellum of speed skaters is larger in the right hemisphere because of their highly developed skills of balance and coordination. Golfers show increased gray matter volume in brain areas having to do with spatial information processing. London cab drivers are well known for the increased size of the hippocampus, reflecting their encyclopedic spatial memory of the city of London.[25] While GPS may have made a dent, London cabbies still say that the knowledge in their minds is far more comprehensive, they don't have to consult a screen, and if suddenly in a traffic jam, they can instantly devise a better route. They can also give you information about sites along the way—restaurants, museums, parks, schools, and pretty much anything else a rider may want to know.

Because making music is a multisensory as well as a motor experience, because musicians have usually begin studying early in life, and because, in highly skilled or professional musicians, nearly all the brain areas that are involved in making music show functional or structural (anatomical) changes, musicians have proved to be of particular interest to researchers studying neuroplasticity.[26] Over the past twenty-five years, musicians' brains have come to be considered models of neuroplasticity.

Functional neuroplasticity

Functional neuroplasticity refers to the brain's ability to change how neurons function. This can either happen as a result of learning, as in learning a musical instrument, or to compensate for damage to a part of the brain. Functional neuroplasticity that occurs because of a sensory deficit, as in congenital blindness, is called cross-modal neuroplasticity and will be discussed in detail in Chapter 7.

During many hours of instrumental or vocal practice, the synapses along the neural pathways become stronger and transmission becomes faster. The behavior of the neurons in those pathways changes. They become "better" at what they do. For example, musicians outperform non-musicians in perception of pitch, timing, and timbre. They also process auditory, motor, and visual information faster and more efficiently.[27] They have superior working memory and are better at musical tasks involving pitch discrimination.[28] Conductors are much better at determining the location of a certain sound source than either other musicians or non-musicians, which is not surprising since conductors must be able to distinguish the sound of adjacent instruments at the back of the orchestra as well as at the front.[29]

You may be thinking, "of course musicians are better at pitch, rhythm, memory, etc. It's natural that years of practice would result in better abilities in those areas." That's exactly the point! Musicians themselves have tended to think about practice mainly in terms of muscles. But the areas of the brain involved in making music change as they practice, neural pathways become stronger, and the enhanced capacity of those brain areas leads to the development of stronger musical skills. Those enhanced brain areas can also have an effect on other cognitive areas (see Chapter 10).

Structural neuroplasticity

Structural neuroplasticity refers to the brain's ability to change its structure as a result of musical training. One of the first studies of structural neuroplasticity compared a group of string players who had played for an average of eleven years with a control group of individuals who had no musical experience. Using magnetoencephalography (MEG), which measures brain activity in milliseconds, they found that the area of the somatosensory cortex corresponding with the fingers of the left hand, used extensively for fingering, was larger in the string players than either their right hands, used to hold the bow, or the control group of non-musicians. The difference was more pronounced in string players who had begun training at an early age.[30] Professional pianists show something called the "omega sign" in the motor cortex of both hemispheres, indicating a larger hand area; violinists show it only in the right hemisphere that controls the left, fingering hand.[31] The brain areas corresponding to the right hand holding the bow don't change.

Studies of neuroplasticity have usually focused on gray matter plasticity. Gray matter refers to the cell body, synapses, and glial cells that support the neuron, all of which look somewhat gray under a microscope. As synapses connect to more neurons and additional neurons are added to the task, there is an increase in gray matter density; this is structural neuroplasticity. In auditory, motor, and *visuospatial* areas of the brain, professional pianists have more gray matter volume than amateur pianists, who, in turn, have more than non-musicians.[32] Visuospatial cognition refers to the ability to shift spatial attention, hold items in visual memory, and detect patterns, all of which are used when sight-reading, which involves high demands on a musician's ability to process complex visual input and connect it with motor output, in real time.

Orchestral musicians have been found to have enhanced gray matter volume in Broca's area, usually linked with speech production but also known to have a role in supporting sight-reading.[33] Musicians have higher gray matter volume in the cerebellum, important for motor learning and cognitive function, with the amount of gray matter correlated to the intensity of practice.[34] They also have more gray matter volume in the hippocampus, important for visuospatial memory.[35] The brain activation pattern for singers is different from other musicians because singers produce and amplify sound entirely within their bodies. Researchers have found that opera singers have increased activation of the primary somatosensory cortex in both the right and left hemispheres representing articulators—tongue, soft palate, lips— and the larynx, all necessary for the production of singing.[36]

White matter, or myelinated axons, also undergoes neuroplastic changes. The corpus callosum, the largest white matter structure in the brain, is larger in musicians than in non-musicians, and larger in musicians who began study before the age of seven than in those who began later.[37] Comparing a group of professional pianists in Sweden with non-musicians, researchers found that increased amounts of practice time in childhood led to greater myelination (white matter plasticity) of axons in the corpus callosum and in the frontal lobe.[38]

In the early years of research, some researchers wondered if some individuals already had larger brain areas that would predispose them to do well studying the piano or another instrument. The answer turned out to be no. Multiple studies monitoring children studying an instrument over a period of years have shown that the structural differences in brain area are clearly a result of practice.[39] And the comparisons we previously looked at showing professional musicians vs. amateur musicians show that the brain changes with practice.

Sometimes brain areas decrease in size

When engaged in various kinds of hand-tapping exercises, professional pianists show far less activity in motor areas than do non-musicians. Extensive training leads to greater efficiency, so not as many neurons are needed to perform routine hand tasks.[40] Structures in the striatum, part of the basal ganglia, assume a kind of executive management oversight, including planning and executing movement and procedural learning (motor skill learning). When first learning an instrument, visual, proprioceptive, and auditory feedback are required, and the striatum has a role in that. But as skill is gained, less feedback is necessary because motor movements become more automated. Fewer neurons are recruited for the same movements, not as much "management" is needed, and the striatal areas decrease in size.[41]

Why does making music drive neuroplasticity in the brain?

Different brain areas show neuroplastic changes in different populations, depending on their activities, as we saw with taxi drivers, golfers, and speed skaters. But it is the intersection of cognitive, sensorimotor, and reward systems that particularly drives neuroplasticity in people who study music.[42] Visual, auditory, and kinesthetic senses combine with motor abilities to allow us to make music—our *sensorimotor* systems. Many *cognitive processes* are also employed in the study of music: thinking, attention, perception, memory, learning, and reasoning. Music activates the *reward circuits* in our brain, and whenever the reward system is activated, it causes us to want more, so we continue making music, driving neuroplasticity. But the neuroplasticity resulting from studying music doesn't just make us better musicians; it has other cognitive benefits as well, as we will see in Chapter 10.

Key Concepts

- Multiple brain areas are involved in the processing of music, including regions in both hemispheres, all four lobes, the brainstem, and the cerebellum.
- The brain changes in response to learning and experience; these changes are called neuroplasticity. Each person's brain develops in a particular way in response to how it is used every day. A musician's brain will develop in a different way from that of an athlete.
- Because music is a multisensory as well as a motor experience, because musicians have usually begun studying early in life, and because nearly all the brain areas involved in making music show neuroplastic changes,

musicians are of particular interest to researchers, and musicians' brains have come to be considered models of neuroplasticity, changing in both structure and function as a result of years of practice.

- All sensory information enters the brain as electrical impulses, and all communication within the brain occurs via electricity and chemicals.
- As we learn music, one neuron connects with another, forming pathways and networks connecting the many areas in the brain involved in making music.
- The intersection of cognitive, sensorimotor, and reward systems particularly drives neuroplasticity in people who study music.

5

Learning and Memory—Two Sides of the Same Coin

The memory is sometimes so retentive, so serviceable, so obedient—
at others, so bewildered and so weak—and at others again, so
tyrannic, so beyond controul!
> —English novelist Jane Austen, *Mansfield Park*[*]

I have many memories connected to my piano lessons at the age of five.
I remember the four blocks I walked to the lesson, the placement of the
piano next to an L-shaped stairway in the music teacher's studio, and
the very large dog that chased me one sunny Saturday morning. But I
have no memories of the lessons themselves, no memories of learning to
match symbols on the page to keys, learning to coordinate my two hands,
or learning how to use the pedal when my legs were finally long enough
to reach the floor. It feels as though I have always known how to play the
piano. Although this may seem strange, it is actually quite normal. I also
have no memory of the steps involved in learning to tie my shoes, ride a bicycle, or drive a car; few people do.

Once the "how" of a complicated motor skill is learned, it becomes automatic, and we have no memory of not being able to do it. This is called *procedural memory*, the kind of long-term memory for knowing how to do things,
for learned motor skills. Once one has learned to play an instrument, no
one thinks about the steps involved, nor are they forgotten. But procedural
memory is only one of many kinds of memory that are a part of learning to
play an instrument and learning music.

The Musical Brain. Lois Svard, Oxford University Press. © Oxford University Press 2023.
DOI: 10.1093/oso/9780197584170.003.0005

Kinds of memory

The brain has so often been compared to a computer that we tend to think of memory for a piece of music as residing in a particular location, like an mp3 file on a computer desktop. Pull it up, click play, and the piece unfolds. But memory isn't found in one place in the brain. There are many kinds of memory; as shown in Figure 5.1, there are several stages in the learning and memory process, and memory is distributed throughout the brain.

Sensory, short-term, and working memory

We perceive the world through our senses. All memory begins as information entering the brain through one of our six senses: taste, smell, vision, hearing, touch, and proprioception (the awareness of the position or movement of the

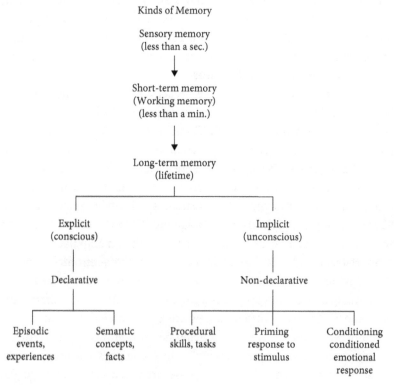

Figure 5.1 Kinds of memory

body). This is *sensory memory* (SM). Whether learning a new piece of music or learning to play an instrument, one *sees* the notation or the instrument, *hears* what the notes or instrument sounds like, *feels* the kind of touch on the keys or strings or notices what it feels like to sing a certain pitch, and is *aware* of where fingers, arms, and body are in space while playing or singing the notes. There is no storage capacity in the brain for sensory memory, so in less than a second, the information that one notices or has paid attention to is sent to *short-term memory* (STM). When learning a new piece of music, it is important to pay attention to as many details in the score as possible because anything you don't specifically notice will not be sent to short-term memory, and that's the first step toward long-term memory.

Close your eyes for a moment and concentrate on all the sounds you can hear. I can hear the grandfather clock chime in another room, traffic sounds on the highway a couple of blocks away, and some kind of power equipment being used by my next-door neighbor. I hear these sounds when I am paying attention. But when I concentrate on thinking about and typing these words, I don't hear any of those sounds. They aren't relevant to what I'm doing so they aren't sent to short-term memory. There is a large intake capacity for sensory memory, the brain can absorb a great deal of sensory information at any one time, but there is no storage capacity. So whatever is not immediately sent to short-term memory disappears.

Short-term memory (STM) is the capacity to hold a limited amount of information for a short period of time, sometimes referred to as a kind of "scratch pad" for temporary recall. Classic examples of short-term memory are recalling a phone number long enough to punch it into the phone or remembering the beginning of this sentence long enough to relate it to what follows. Short-term memory is sometimes used synonymously with *working memory* (WM), although some researchers still differentiate the terms— STM referring only to short-term storage and WM to holding information in your mind so that it can be mentally worked with or manipulated.

Short-term and working memory rely on the prefrontal cortex, the "thinking" area of the brain where executive functions are regulated.

Thinking about sight-reading is a good way to contrast the different kinds of memory. As you sight-read a new score, you are always scanning a bit ahead of what you are playing, noticing notes, rhythms, patterns, dynamic markings, etc. Your brain is taking in information via sensory memory. What you have specifically noticed passes from sensory memory to short-term memory. Meanwhile, working memory keeps that information in mind long enough for your brain to map out how to play it. That involves relating patterns you see to patterns you already know that are in your long-term memory—making them easier to play. If you sight-read frequently, you will be better at it because a lot of patterns you encounter will already be in long-term memory.

Short-term memory has a small storage capacity and a short duration, less than a minute. The capacity for short-term memory is usually thought to be seven items of information, plus or minus two. But the capacity of short-term memory (or working memory) can be increased by "chunking" or grouping smaller individual pieces of information into larger patterns, basically recoding the information in the brain. For example, the eight digits in the number 10061975 are near the limit of what the brain can hold in STM. But if thought of in 3 chunks—month, day, year (10-06-1975)—rather than eight individual digits, it is easier to remember all eight digits, and some STM space has been freed up for more numbers. Similarly, one might think of G, B, D, F as one chunk, a dominant seventh chord on G, rather than four individual pitches, leaving room for five or six additional chunks of information in short-term memory.

Chunking is based on information already known and stored in long-term memory. One must know that dates can be expressed as 2 + 2 + 4 digits to chunk eight digits as a date, and one must have some knowledge of music theory to be aware that the pitches G, B, D, F are a dominant seventh chord. The more knowledge one has about a subject, the more readily information can be chunked, which is why those who sight-read a great deal are so good at it. They are accustomed to seeing patterns and chunking musical information.

To retain short-term memory, we must do something to facilitate its transfer to long-term memory, and for musicians, that means practice. Do nothing and short-term memory disappears.

Long-term memory

All memory other than sensory, short-term, and working memory falls under the category of long-term memory, and that is the goal when learning music. But long-term memory isn't a single entity. There are two long-term memory systems, and they rely on different brain networks. Procedural memory, sometimes called implicit or non-declarative memory, is concerned with forms of memory that are not dependent on conscious processes, like riding a bike. Declarative memory, also called explicit memory, is concerned with conscious memory or memory that can be verbally stated as facts or ideas. Memory for a specific piece of music is declarative memory.

Procedural or implicit memory

Procedural memory is the memory for a motor skill such as tying shoes, driving a car, or playing a musical instrument. It is acquired through repetition, through trial and error, and once learned, is so deeply embedded that it is not forgotten. Procedural memories are difficult to explain, much easier to demonstrate. Most men, if asked how to tie a tie, would find it difficult to explain and would prefer to show you. We teach children to tie their shoes by demonstrating, not by explanation. We do the same thing when teaching music. It is easier to show someone proper bow placement on the cello than to explain, easier to demonstrate hand position at the keyboard than to give directions. Procedural memory is a kind of *implicit* memory. Other kinds of implicit memory include habits, conditioning, and preferences, all of which are unconscious, like procedural memory, but they have their origin in our past experiences or activities and were acquired without our noticing or thinking about them. We usually are unaware as to how we acquired certain habits. They haven't been practiced and are not accessed through conscious thought. Implicit memories are part of who we are, and they influence our behavior.

In order to learn to play a musical instrument, the brain must acquire and implement what neurologist Alvaro Pascual-Leone calls a "translation mechanism" to convert knowledge into action.[1] Many people who begin lessons have seen someone play an instrument and have fallen in love with it. They may have a general knowledge of how the instrument sounds, how it is held, how one sits at it. But when one begins to learn the instrument, movements are tentative and slow. Each movement has to be thought through very carefully. Although bimanual coordination is used for everything we do, coordination of both hands somehow seems very difficult when we begin to learn many instruments. This is because the two hands will no doubt each be performing different movements, and the brain prefers limbs to perform synchronous, symmetrical movements like walking, riding a bicycle, or swimming, not different movements in each hand.[2] Everyone is familiar with the classic "coordination teaser" of rubbing your stomach while patting your head. Here's another: lift your right foot and do clockwise circles in the air. While doing that, draw a number 6 in the air with your right hand. You may suddenly find your right foot going counterclockwise to match the counterclockwise movement of the right hand. The brain would prefer that both right hand and right foot go in the same direction.

Overcoming the brain's preference for symmetrical movement is necessary to do the teasers—or to play a musical instrument. Think about a violinist drawing a bow across the strings laterally with the right hand while the fingers on the left hand, which is held out in front of the violinist, must move up and down on the fingerboard while at the same time moving back and forth to find the exact placement for a specific pitch. Playing a string instrument may be the ultimate coordination challenge, but playing any other instrument presents its own coordination challenges.

The motor cortex is involved in movement, but several other areas of the brain are involved as well in the bimanual coordination necessary for playing an instrument. These include the cerebellum, the supplementary and premotor areas, the cingulate motor cortex, the corpus callosum, which carries information between the two hemispheres, and the basal ganglia, which not only facilitates voluntary movement but helps to inhibit competing movements.[3] Each has a specific role in coordinating and integrating both of our limbs into a sequence of muscle movements while suppressing mirror movements. As we practice, we imprint new patterns and suppress the brain's natural inclinations for symmetry, and neuroplasticity strengthens these movement patterns and makes it possible to use each of our limbs in the different ways required for our instrument.

Musicians learn to coordinate what is seen on a page of notation with the movements necessary to make the sounds the notation represents. Sensory memory takes in information as we match movement to sound, transfers that to short-term memory, and eventually, over time, the skill enters long-term procedural memory. Our brains translate what is known about playing the instrument into being able to do it—knowledge into action, Pascual-Leone's "translation mechanism." After an instrument is learned, the struggles to master the coordination problems tend to be forgotten.

As skill level increases, practice is no longer about learning to coordinate our limbs enough to play the instrument and match sound to movement, but about learning the specific technical and movement challenges that arise in each new piece of music. Learning *how* to play the instrument has entered procedural memory, the memory for how to do something (although singers say theirs is a constantly evolving instrument and one continues to be aware of the coordination necessary throughout the body). The two brain areas involved in procedural memory are the basal ganglia and the cerebellum.

The basal ganglia, lying deep within the brain, are involved in the formation of motor programs, in packaging smaller movements into larger ones, and in coordinating sequences of motor patterns used to play an instrument so they don't constantly have to be relearned each time they are encountered. The movements become routine, part of procedural memory.

The cerebellum is important for the timing, execution, and coordination of the movements. It's important for the fine motor control needed to play an instrument. Another structure, the amygdala, can also be involved, as we will see later.

Declarative or explicit memory

When we use the general term "memory," we're usually referring to declarative memory, the conscious recollection of facts, previous experiences, and concepts, also called *explicit* memory because these memories can be verbalized. When we worry about memory before a performance, it's not procedural memory that we are worried about—no one forgets *how* to play an instrument or sing. The concern is about forgetting some detail of the music: chord changes, how the theme is different the second time, or the text for a song. That's declarative memory, and there are two kinds: *semantic* and *episodic*.

Semantic memory refers to factual knowledge, to *knowing*. When we first learn an instrument, we concentrate on pitches and rhythm. As our skills and knowledge become more advanced, we add theoretical, historical, and musical information to the complex framework that constitutes our semantic memory. That is what we draw upon as we learn a new piece of music. Semantic memory is *knowing* facts and details about a piece of music—being able to name notes, talk about key relationships, cite information about the composer, describe the structure. A singer needs to be able to recite the text; in an opera, a singer must know the translations not only for her own role, but the roles of the other singers as well, and stage directions must also be memorized, all of which is semantic memory. Semantic memory is also involved in being able to transfer concepts from one piece of music to another, building knowledge over time.

Episodic memory is *remembering* events or experiences. We remember important occasions such as commencements or weddings. We may remember our first recital, our piano lessons with our college professor, a performance that went stunningly well, or one that we thought was a disaster. We remember where these events took place, the time of day, and how we felt. When we reminisce about events in the past, we are using episodic memory. Episodic memory provides us with a sense of our own personal history.

The hippocampus is the area of the brain responsible for converting short-term memory for a piece of music into long-term memory, and this happens over time with practice. After the memory is consolidated, the hippocampus sends the various elements of the memory back to the cortex where they were first processed—visual, auditory, somatosensory, and motor—and that's where they are stored. The hippocampus is also important for processing and storage of spatial memory. For example, a pianist must know exactly how far to stretch a hand to play an octave. A string player must know exactly where first position is on the fingerboard. These skills require spatial memory and that is stored in the hippocampus. Brain areas where explicit and implicit memory are stored can be seen in Figure 5.2.

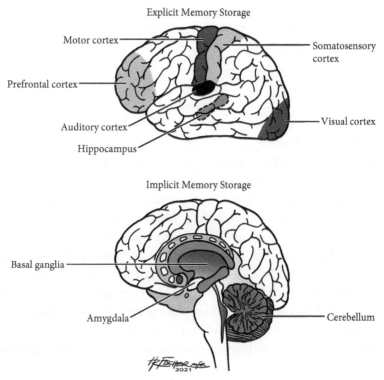

Figure 5.2 Explicit and implicit memory storage

While semantic memory is more important for memorizing all the details about the piece, episodic memory also has a significant impact because it links emotions with memory, and a structure called the amygdala has a role in that, as we will see later.

Procedural (implicit) memory is used to play our instrument.
Declarative (explicit) memory is used to play a specific piece of music.

Stages of learning and memory

Many neuroscientists say that learning and memory are two sides of the same coin. Learning is the process by which new information is encoded in the brain, and memory is the encoding of the information itself. It is no surprise that how well a piece of music is learned determines how easily it will be recalled from memory. Still, the two processes tend to be dissociated. Many students talk about "learning" a score and then "memorizing" it, as though memory is an add-on. Memory, however, is inextricably intertwined with the learning process. There are five stages of memory that are important to know when learning music.

Encoding

Encoding is the first step in learning or memorizing music. When sight-reading a new piece of music, a great deal of sensory information is encountered—rhythmic or pitch details seen in the score, the sound of the melody, how it feels when the keys or strings are touched. A meaningful neural representation of the piece is formed in the brain by neural pathways connecting sensory areas as well as the motor areas necessary to make the movements to produce the sound. This memory trace contains those elements of the piece that we noticed, paid attention to, and entered short-term memory. This is encoding—the initial representation of the piece. If the piece is abandoned and not practiced, this memory trace will fade and disappear. For a piece to become part of long-term memory, the memory trace or neural representation of the piece must be consolidated. This happens with practice.

Consolidation

As we practice, the brain consolidates the initial memory traces or neural pathways—strengthening, reorganizing, stabilizing, adding meaning, filling in blank spots, making connections to previous knowledge. Information is added that we hadn't noticed when we first sight-read the piece. More practice means paying more attention. Incorrect rhythms or pitches are fixed; we notice rhythmic, melodic, and harmonic patterns; we figure out the best fingering or bowing, learn the structure, solve technical problems, and think about various interpretations.

Many people think of this simply as practicing, not aware that it is a biological process in the brain. Neural pathways become stronger as we practice the music. Connections at the synapses are strengthened, more neurons are recruited as new information is added. New neural pathways are formed, and some are re-routed. Transmission of the information becomes faster as more myelin is wrapped around the axons. "Cells that fire together, wire together." As one practices and continues to adjust and refine, the neural pathways are reinforced, and the consolidation of memory becomes stronger. Consolidation also occurs during sleep, as we will see in the next chapter.

Storage

Perhaps the idea that there is a specific place in the brain where a particular memory is stored stems from the term itself. We tend to associate the word "storage" with a place to put things, whether an attic, basement, storage facility, or an external hard drive for a computer. But the two kinds of long-term memory, procedural and declarative, are stored in different areas in the brain, as we have seen. Declarative memory itself is stored in multiple brain areas.

We can't directly control where the hippocampus sends information for storage. What can be controlled, however, is what we pay attention to during practice because that determines what information, and how much information, is encoded and how securely it becomes consolidated and stored in long-term memory. Anything visual that we have paid attention to will be stored in the visual cortex, auditory in the auditory cortex, etc. Information won't be stored if it hasn't been focused on during the encoding and consolidation process.

Retrieval

Retrieval is the stage of the memory process that is relied upon in performance. During retrieval, the brain reconstructs the various elements of the piece—visual, auditory, kinesthetic, motor—stored throughout the brain and linked by neural networks. Playing a piece of music from memory requires revisiting the neural pathways that were formed when the brain encoded and consolidated the memory, and how readily the piece is remembered depends on the strength of those pathways.

Reconsolidation

Memories do not remain the same over time. The very act of retrieving the memory of a piece changes it and this is called reconsolidation. Neuroscientist Joseph LeDoux comments that "the brain that does the remembering is not the brain that formed the initial memory. For the old memory to make sense in the current brain, the memory must be updated."[4] Each time a particular piece is performed from memory, additional information is brought to it that has been learned since the initial memory was formed. A musician rarely plays a piece of music the same way twice. Arthur Rubinstein once commented that he was fond of changing a fingering during performance. Most musicians wouldn't suddenly change fingerings or bowings in performance, but the longer a piece is known and performed, the deeper one's understanding of the piece becomes, and the conception of the piece may change. Dynamics may change, more liberties may be taken with tempo, more emotional depth may be developed, one may practice new fingerings or bowings, technique becomes more secure, and as more experience is gained in front of an audience, one's comfort level grows. The memory itself changes, and as it does, it reconsolidates and is again stored. So while it may feel as though it's the same piece being recalled, the memory has changed based on the new information, however subtle, that is brought to it.

When emotion meets memory

The two memory systems, one involving the hippocampus and supporting declarative memory, and the other, involving the basal ganglia and cerebellum and supporting procedural memory, are independent, though they act together when emotion meets memory. The brain structure common to both kinds of memory is the amygdala.

The amygdala, part of the limbic system, is involved in the processing of emotions. It is essential not only to our ability to feel emotions but to perceive them in other people. The amygdala is involved in the consolidation of declarative memories that have strong emotions, whether that emotion is positive or negative. Attention is important for the initial encoding of a memory, and emotion captures attention, whether it is positive or negative.[5] Music teachers have always noticed that students will learn a piece much more quickly when they feel an emotional connection to that piece. Emotional experiences are "tagged" by the amygdala as important and that facilitates consolidation by the hippocampus. In *This Is Your Brain on Music*, neuroscientist Daniel Levitin points out that the teen years are very emotionally charged, and music experienced in those years tends to be remembered because the amygdala has "tagged" the memories associated with that music as important.[6] Other emotionally charged times in our lives—weddings, funerals, commencements, or performances that we deemed disasters—are similarly "tagged" by the amygdala and become unforgettable.

The amygdala is particularly known for its role in the processing of fear, for triggering a heightened fear response usually referred to as the "fight or flight" response. Although the fight or flight response evolved in our prehistoric ancestors to protect them from the danger of wild animals or enemies, today it is more likely to be a response to heightened stress or anxiety such as that precipitated by a performance. I mentioned earlier being chased by a large dog at the age of five. I have a conscious, declarative memory of the event itself, of the large dog in front of me, exactly where I was, and the color of the dog. I remember the fear, although I don't experience the fear now when recalling the memory. This is declarative memory, and I remember the event because my amygdala tagged it as important at the time.

On the other hand, I have carried my childhood fear of dogs into adulthood. The fight or flight response that was triggered at the time became part of my implicit memory of dogs. It's not conscious and I can't will it away. You may recall that implicit memory does not just include procedural memory for playing an instrument, it also includes conditioning. My dog experience conditioned me to fear dogs. The same thing happens with performance. If something upsetting happens during a performance, such as a memory lapse, the negative emotions experienced at the time are incorporated both into declarative memory and implicit memory. When recalling the event (declarative memory), the memory lapse or mangled technique looms large. The same feeling of devastation isn't experienced with the recollection, it is simply remembered. At the same time, those negative emotions were reconsolidated into implicit memory of performance, or of that particular piece, raising the specter of performance anxiety about the same thing happening again.

However, just because a performance is recalled as being calamitous doesn't mean the memory is accurate, just as my memory of size of the dog or how menacing he was may be inaccurate.[7] The vividness of negative emotions about an event often overrides accurate details of the event itself. One minor memory lapse in an otherwise wonderful performance can color the overriding experiential memory as a negative one.

Is there a solution? Yes, and it involves the prefrontal cortex, the thinking part of the brain, which is also involved in performance and memory, but becomes overridden by the emotional part of the memory. There are various current research investigations looking at how the prefrontal cortex can inhibit the emotional reaction of the amygdala to reverse fear conditioning. But there are also practical mind-body techniques to break the cycle of one bad performance leading to another due to performance anxiety. Two excellent books that address anxiety in performance from the perspective of the mind and body are Julie Jaffee Nagel's *Managing Stage Fright* and Vanessa Cornett's *The Mindful Musician*.[8] It is important to remember, however, that addressing performance anxiety is a matter of rewiring the neural circuits that have attached fear to the memory of performance. It takes time and multiple repetitions of positive performance experiences to rewire the brain. That often begins with small, less threatening performances, perhaps for friends, and gradually working up to more significant performances, replacing the emotion of fear attached to memory of performance with more positive emotions.

What happens when we walk onstage to perform?

When a performer walks out onstage ready to perform, two memory systems are at work in the brain, the procedural system for playing our instrument or singing, and the declarative system to remember the piece of music. Daniel Levitin says that when one sits down to play a piece of music, the brain must execute what is known as a "motor-action plan."[9] When a piece is initially learned, smaller elements of motor movements are first learned that then combine to form larger sequences of movements, facilitated by the basal ganglia—movements for motives lead to phrases, and then to sections, etc. The sequence of movements must happen in a particular order for a particular piece of music—motor movement following motor movement, pattern following pattern, becoming the motor-action plan for the piece. The brain is directing movement patterns, not individual muscles. With practice, the neural pathways are strengthened. Fingers eventually seem to move on their own, but the signals are coming from the brain. Once a piece is learned, the motor-action plan becomes part of procedural memory, the "how" of playing a particular piece of music.

A few years ago, I was introduced to domino art, and there are interesting parallels between domino art and a "motor-action plan." For those unfamiliar with this art form, it is the construction of elaborate arrangements of thousands of colored domino tiles that the artists then knock down in artistic chain reactions. Many of these artists have their own YouTube channels where the building and the knock-down are documented. The creations may be abstract, a portrait, company logo, or replicas of iconic paintings. *[My personal favorite, seen on the companion website as item 5.1, is a re-creation of Van Gogh's "Starry Night" by Canadian domino artist Flippy Cat ⏵.]*

You might imagine that with tens, sometimes hundreds of thousands of dominoes, there could be occasional problems with the knock-down, and in fact, there are. Flippy Cat says that the chain reaction doesn't always go as planned. A crucial domino may be inadvertently left out or incorrectly placed, something may fall from an overhead fixture or camera filming the event, and the chain reaction stops midstream.[10] This is not unlike an unexpected event occurring during a performance that causes our motor-action plan to be derailed. A distraction, perhaps a noise in the hall, a chord that is reached for and missed, the mind flashing to mistakes we may make, and suddenly, what comes next is forgotten and the motor-action plan stops.

Many people studying music have been told that muscle memory or motor memory is unreliable and that it is the first to fail in a stressful situation. That's not the case. The motor-action plan may stop, along with the performance, but it isn't the motor memory for the piece that has failed. No one forgets *how* to play or sing the piece. The brain knows the sequence of motor movements in the serial order that determines the piece, and if shown the score, the performer can continue. I once heard a well-known pianist in recital come to a complete stop in the middle of the first page of a Beethoven sonata. He started again from the beginning, came to the same place and stopped again. Finally, he got up, left the hall, and after a few minutes returned with the score. He set it on the rack, began again and played the sonata exquisitely without once turning a page of music. His motor memory, or motor-action plan for the piece, was completely intact. What he had forgotten was some detail of declarative memory that would have allowed him to re-enter the motor action plan at some point other than the beginning of the piece. That is true for everyone. It may sound as though the motor memory or procedural memory has failed because of a stumble or stop, but it's some detail of declarative memory that is blocked or forgotten, stopping the performance.

Unless gifted with an eidetic, or photographic memory, most musicians are not able to begin on any note in a piece. But multiple points of entry to the motor-action plan, other than the beginning of the piece, are essential. Learning these points of entry or retrieval cues is part of the encoding and consolidation phase of memory, and part of successful practice (see Chapter 6).

What constitutes successful practice?

Cognitive psychologist Roger Chaffin studies musicians and memory. He has tracked several professional musicians from the time they first begin practicing a new work until they perform it. His interest has been in the psychological processes involved and the musical judgments made by the musicians while practicing that ultimately shape the performance. He has studied a pianist, a singer and conductor, a jazz pianist, and in the most extensive study, a cellist whom he followed through the process of learning J. S. Bach's Suite No. 6 for Solo Cello and continued to follow for a period of three years and ten public performances of the work.[11]

He found that professional musicians playing from memory rely on the same principles as those used by experts who rely on memory in other disciplines: (1) meaningful encoding of the musical material; (2) use of a well-learned retrieval structure to access the material; and (3) practice to ensure rapid retrieval from long-term memory.[12]

What is "meaningful encoding of the musical material?" When practicing, motor movements necessary to play the piece are encoded and consolidated and a motor-action plan is learned, leading to procedural memory. One may draw on chunks of musical expertise already in long-term memory, such as scales, chords, and patterns. But if one practices only motor movements, the piece hasn't really been learned. All the musical facts about the piece need to be encoded as well. What is the structure? Where are the key changes? What are the notes in a series of complex chords? Where do rhythmic changes occur? Where is an unusual fingering needed? This is declarative memory—the facts and details about the piece, and the more facts, the more secure the memory will be.

Chaffin's second principle is the "use of a well-learned retrieval structure to access the material." What happens if a performer has a memory slip in performance? Start again from the beginning? If so, what happens when the same place is approached the second time? If the performer's only access point is the beginning, then there isn't enough encoded information—there is no retrieval scheme. Remember, once the technical difficulties of a piece are mastered, you have a motor-action plan, but it can be derailed when some detail of the music is forgotten. Multiple points of entry are required to be able to access the motor-action plan. The structure of a piece provides a ready-made retrieval scheme. One can pay attention to and encode into long-term memory the beginnings of the exposition, development, and recapitulation in a sonata, or one can encode into long-term memory how the ABA sections begin in a piece, or how the coda begins. With complex new music, one can encode the structural sections of the work. Knowing the structure of a piece provides access to chunks of information in long-term memory, allowing one to get back on track when something unexpected happens.

While the structure of a piece remains the same for everyone, other re-trieval cues may vary. One person may use texture changes as retrieval cues, another may use difficult fingering patterns, chord progressions, technical difficulties, melodic or rhythmic patterns, or phrasing. Use what makes the most sense, or perhaps what has been the most difficult to learn. But have multiple retrieval cues (sometimes called performance cues or landmarks) to provide access to the motor-action plan for the piece.

The retrieval cues may vary, but what is crucial is Chaffin's third point: prac-tice to ensure rapid retrieval from long-term memory. What often happens in a memory slip is the inability to access declarative memory quickly enough to realize where you are and to get back on track. Motor memory is moving faster than declarative memory, or fingers are moving faster than the mind. The performance flounders until some landmark in the piece can be remembered.

One needs to be able to access retrieval cues *as fast as* one is playing the piece in order to feel secure about memory. That means identifying a set of cues or markers in the piece that are meaningful, practicing starting at each of these places, and then practicing accessing each one at the same speed as one is playing, so that when, not if, the unexpected happens, you always know where you are (see Chapter 6).

This doesn't diminish what some people refer to as the "flow state," that state of being totally immersed in the music during performance. In fact, having ready access to the declarative memory of the piece makes the flow state more possible, because encoding and consolidating factual details in memory simultaneously with auditory and motor memory gives a great deal more freedom in performance.

Some say that if you really focus and know the piece, you won't lose your train of thought and have a memory slip. But life isn't predictable. Playing or singing in a large performance space is not the same as playing or singing in your living room or in your teacher's studio. A door may slam, someone coughs, a cell phone rings. A friend of mine experienced a power outage and blackout, but he continued playing the Beethoven sonata to the end of the movement. A dog once wandered onstage while I was performing (yes, me and dogs). Anything can occur to startle a performer into losing focus or train of thought, and anyone can be startled. *That is not a failing*, it's a part of being human. But if one has instant access to declarative memory through retrieval cues, the motor-action plan can be restarted—usually without an audience being aware that anything has happened.

The goal during practice is not the same as the goal in performance. One practices to learn the material; one performs, with or without a score, to communicate with an audience. Sergei Rachmaninoff once said that in every piece we choose to learn, there is something—a chord sequence, a haunting melody, a harmonic progression—that we fall in love with on first read-through or first hearing. And that is the reason we choose to learn that piece. He went on to say that the trick in performance is to play it as though you are discovering that chord, melody, or progression for the first time. This is difficult because so many hours are spent rehearsing a piece that we may lose sight of why we wanted to learn it in the first place. But with the right kind of practice, the mechanics of performance can become so automatic that there is freedom to concentrate on the emotional expressiveness, on the "redis-covery" of the reasons one loves this particular piece of music and chose to learn it.

Key Concepts

- Learning and memory are two sides of the same coin: learning is the process by which new information is encoded in the brain; memory is the encoding of the information itself.
- The three categories of memory are sensory, short-term memory (including working memory), and long-term memory. All are used in learning our instrument and learning and memorizing music.
- The two kinds of long-term memory are declarative and procedural. Procedural memory is necessary for playing our instrument. Declarative memory is used to play a piece of music.
- The five stages of memory are encoding, consolidation, storage, retrieval, and reconsolidation, and all are vital in learning and memorizing music.
- When one has a memory slip, it is not procedural memory, or motor memory, that is lost; it is declarative memory, the facts and details of the piece.
- Successful practice includes meaningful encoding of the musical material, use of a retrieval structure to access the material, and practice to ensure rapid retrieval from long-term memory.

6

Practice—It's All about Quality

The good practicer tastes the vitality of adventure and the dramatic
rewards of risk-taking.

—pianist William Westney[*]

Musicians worry a lot about memory—about memory slips in performance, or why a particular piece of music is so difficult to memorize, or why
something learned in middle school is easier to remember than something
memorized last month. If learning and memory are two sides of the same
coin, as we saw in the previous chapter, what kind of practice, or learning,
will ensure secure memory of the musical material so that performance feels
more comfortable?

Why forgetting improves memory

No one wants to forget the music in the middle of a performance but
forgetting while in the process of learning is actually beneficial, according
to psychologists Robert Bjork and Elizabeth Ligon Bjork. The Bjorks have
studied learning and memory for over thirty years, and they have proposed
that forgetting is a necessary part of the learning process. In fact, their research lab at the University of California, Los Angeles, is called the Learning
and Forgetting Lab. There are several principles that the Bjorks identify in
their work that have particular relevance to learning and memorizing music.[1]

Storage strength is different from retrieval strength

The brain has unlimited capacity for storage of memory, but very limited capacity for retrieval. *Storage strength*, which accumulates as we practice, refers

The Musical Brain. Lois Svard, Oxford University Press. © Oxford University Press 2023.
DOI: 10.1093/oso/9780197584170.003.0006

to how well the details for a piece of music are encoded and consolidated. How many details have been committed to memory? How many ways has the material been learned? The more senses we use to encode and consolidate information—visual, auditory, kinesthetic, proprioceptive—the more storage will accumulate. As facts and details are encoded and consolidated, neurons are added to the network, and the connection at the synapses between neurons along the neural pathways is strengthened. Having a great deal of storage strength, however, does not necessarily mean that a memory is effortlessly retrievable. That is dependent on retrieval strength.

Retrieval strength refers to how accessible that piece of music is right now. After endless repetitions in the practice room, can we play it from memory in a teacher's studio or onstage? Have retrieval cues been learned and how quickly can they be accessed? In the middle of a memory slip, how quickly can the motor-action plan (discussed in previous chapter) be accessed? With multiple ways to access the piece, multiple places to start "cold" without looking at the score, retrieval strength is strong. If the motor-action plan can be retrieved at the same speed one is playing the piece, retrieval strength is strong.

Theory of disuse

Memory for a piece of music doesn't disappear over time—it just becomes inaccessible. Since the brain has an unlimited capacity for memory storage, there is likely to be a great deal of music stored in your brain if you have studied for a long time. However, it isn't easily accessible because of the limited capacity for retrieval. Pieces one learned for grade school recitals are still there but inaccessible, and a look at the score would quickly return them to accessible memory. But no one needs childhood pieces to be accessible, as they would take up too much of the limited retrieval space that one needs for current music. Music you don't currently need is forgotten and, according to the Bjorks, this is the *theory of disuse*. Only the most relevant information is available, not everything one ever learned.

The role of forgetting

Forgetting doesn't undo learning, it creates the opportunity for more learning. If retrieval strength of a memory is high, room isn't available for further learning. Being able to access a piece by memory from multiple places in the score requires high retrieval strength and lowers the possibility of adding new details. Forgetting, however, creates space and opportunity to add new information, strengthen retrieval cues, reinforce encoded pathways, and increase storage.

Desirable difficulties

The Bjorks call a learning strategy that requires effort a *desirable difficulty*.[2] The more work required to retrieve a memory, the stronger it becomes—both storage and retrieval strength are increased. Certain kinds of practice present desirable difficulties, forcing learning to slow down, creating the opportunity to add relevant information, and presenting more opportunities for the neural pathways to be strengthened. Thus, long-term retention becomes stronger. It may seem counterintuitive, but it will become clearer when we explore the practice strategies in this chapter.

Most people tend to think that if the way they are studying or practicing feels good, they must be learning a lot. The opposite is true. The kinds of practice that lead to secure learning don't feel good. They can be frustrating, slow, and often leave the impression that one isn't accomplishing very much. But is the purpose of practice to feel good or to learn the music securely? There is a difference between practicing to learn motor skills and practicing for performance. Practicing motor skills takes a lot of repetition as one perfects technique and creates motor action plans. Practicing for performance requires a different kind of practice, the practice of declarative memory.

The myth of repetition

Cognitive and educational psychologists have been studying learning for over 100 years—what is effective and what isn't. A few years ago, a group of researchers from several universities reviewed more than 700 scientific articles about the ten most common learning techniques to determine which were most effective.[3] While most of those studies looked at academic learning, the principles apply to music as well. The clear loser in this review of learning techniques was repetition.

Perhaps because instruments are learned through repetition, there is a tendency to think that a piece of music should be learned the same way. Many people are convinced that repetition, or what researchers call massed practice, is the best way to learn new information, whether it's the conjugation of French verbs, the important battles of World War II, or a Chopin prelude. Studies looking at student music practice have shown that the technique used most often is repetition.[4] Many teachers at all levels encourage students to repeat a passage or excerpt until they get it right. I had a teacher when I was eight or nine who wanted a section of a piece played "five times perfectly" before moving on. The assumption was that the section was then learned. Not true, as it turns out.

In sports, continuous repetition is called *blocked practice* and it refers to practicing a particular skill over and over, and then another. A tennis player may do a "block" of repetitions of backhands, then move on to a block of serves, then on to follow-throughs. But whether it is called massed or blocked practice or simply repetition, it doesn't work. One research study after another has found that it is the least productive way to practice. Yet even if one is made aware of this, repetitive practice maintains a strong allure because it *feels* as though progress is being made. Spend a couple of hours on a short excerpt of music and, in all likelihood, it will sound much better at the end of that time. Motor skills will have improved, fluency has increased, and that is often equated with "learning." But actual learning of the musical material, the details of the piece, will not have improved because with each repetition, the material is being recalled from short-term memory. It's no different from repeating a phone number over and over until it is used a minute or two later, and then promptly forgotten. We learn rapidly through repetition, but we also forget rapidly. We don't really know how well we know a piece of music until hours or days later when it must be retrieved from long-term memory.

Obviously, it is necessary to spend tens to hundreds of hours on a piece—that's how the motor skills necessary to play it are acquired. But having the motor skills to play a piece is quite different from the kind of memory necessary to perform the piece. What is mastered through repetition is *procedural memory*—the "how to" of the motor skills. But for performance, we also need *declarative memory*, the memory for facts and details. And there are far better ways to practice solidifying declarative memory than endless repetition.

The two best strategies for learning and memory

There were two clear winners in the study mentioned earlier to determine best learning techniques: *Practice testing* and *distributed practice.*

Practice testing

Practice testing is about practicing retrieval from long-term memory. We don't usually test memory until close to a performance, wanting to have the piece solidly memorized before trying it out for anyone. The problem is, there is no way to know how good your memory is until you add the stress of playing for someone. Practice testing does two things: (1) when you retrieve a memory for a piece of music by playing it, the memory is strengthened through reconsolidation; and (2) practice testing is diagnostic, revealing what you don't know. The earlier the learning process is given a test run, the sooner gaps in knowledge become apparent and can be corrected. When learning a new piece, it is a good idea to attempt to play it from memory as soon as a short section is learned. As soon as retrieval cues are identified, begin from memory at each of those cues.

Even children can do this on a small scale. Before they learn that memory is supposed to be difficult, they will consider it a game to be able to begin at multiple points in the piece. Mistakes are good because they force one to think about and fix the mistake—the role of forgetting. Unexpected problems sometimes come up in performance that never happened in rehearsal, and we say, "Wow! I never would have expected to run into trouble in that spot." But if memory is tested often, all those potential problem spots will be found, and they can be addressed *before* a performance. The brain will have to make the distinction between what is correct and what isn't, and the correct information becomes encoded more securely. Retrieval strengthens the memory.

Practicing memory retrieval so early in the learning process will be frustrating, but that's the concept of *desirable difficulties.* The more one struggles during initial attempts to play from memory, the more one learns about what isn't known. As notes and rhythms that were previously uncertain now become consolidated, the neural circuits become stronger.

An effective way to do practice testing is to number the retrieval cues for a particular piece of music and write the numbers on small slips of paper, or if retrieval cues are sections of a piece, label them as exposition, development, or whatever the sections are in a particular piece. Place the labeled slips of paper in a box and pull out one at a time. If you can't instantly begin the piece at that cue, you probably don't know the piece as well as you thought. Or frequently go back to something you were practicing fifteen minutes ago, or an hour ago, or a day ago, and see if you can play it memorized. First thing in the morning, try to play from memory something you worked on the day before. A friend says that she developed absolute pitch (perfect pitch, see Chapter 3) because at the age of five, her piano teacher told her to think or sing a pitch each morning and then go to the piano and see if she was right. She was developing pitch memory, which is what absolute pitch is. Trying out a piece first thing in the morning does the same thing for memory of a piece of music.

There are many ways to do practice testing of memory, and you will think of many once you begin the process. Singers can speak the words faster than the tempo of the piece. They can write out the words they remember, leave a blank where they can't remember a word or phrase, then go back later to check for the correct words and solidify the memory. Various kinds of practice testing may seem like mechanistic and unmusical ways to practice. However, learning the material so well with memory testing gives you a sense of freedom in performance and allows deeper concentration on communicating the music's emotional content.

Distributed or spaced practice

Spaced or distributed practice has been studied more than any other kind of learning, so you would think that it would be very familiar by now, but it's not. Hermann Ebbinghaus wrote about the spacing effect (distributed practice) in 1885.[5] He studied lists of nonsense syllables and tested what he could remember. He discovered that more information entered long-term memory when the study sessions were spaced out, rather than trying to cram learning into one session. Nearly 100 years later in 1978, Alan Baddeley confirmed the Ebbinghaus theory. Baddeley compared massed learning (repetition) with spaced learning with groups of postmen learning to type alphanumeric code material and found that, although the material seemed to be learned faster in massed practice, it was retained longer when learned over a period of days.[6]

Spaced or distributed practice presents another kind of desirable difficulty. The more study sessions are spaced apart, the more effort is required for remembering, and that leads to better long-term memory. Yes, musicians already space out practice over days, weeks, even months to learn a difficult work, but the concept of spaced or distributed practice is a bit different. This idea suggests practicing a piece or excerpt for a shorter amount of time than usual, and then leave it to do something else. Come back to it later—initially maybe thirty minutes later, adding time to the spacing effect until not returning to that excerpt for a day or two.

There are a couple of reasons that this works. The first is that the brain likes novelty and pays attention to what is new. The more an excerpt is repeated during practice, the less novelty it has. The brain stops paying attention and all those repetitions really don't do any good. Spaced practice means spending a shorter amount of time practicing a particular excerpt and stopping before it feels as though the practice is finished. After a half hour of doing something else, the excerpt will feel fresh again, the brain will pay more attention, and it will notice more information to encode in the neural network.

The second reason spaced practice works is that, like practice testing, it is difficult and feels as though it isn't working. But spaced practice makes the brain work harder. The more effort is expended, the more successful learning becomes—desirable difficulties. If a particular piece of information can't be recalled—whether the interval is a fifth or an octave, whether the melody begins on the second or third beat—it must be looked up and the brain adds and consolidates more information, filling in blank spots, making the synapses in the neural pathway stronger. With spaced practice, the brain makes stronger connections.

Other excellent practice strategies

Interleaved practice

Interleaving is related to spaced practice, but instead of having a cup of coffee during the spacing interval, you practice different music. Interleaving means practicing multiple pieces of music concurrently and alternating among them. Mix up, or interleave, pieces or excerpts of pieces with different technical or musical challenges. This creates what is called the *contextual interference effect* (CI). The interference created by switching from one piece to another leads to poorer performance in the short term but produces far

superior retention compared with repetitive practice.[7] As with spaced practice, practice each excerpt or piece for a short amount of time, leave it before
fully accomplishing the task intended, and go on to something else offering
a different musical or technical challenge. It feels counterintuitive and not
productive, but it demonstrates the difference between learning and performance. With repetitive practice, performance may be great in the short term,
but real learning hasn't happened because there is no retention. On the other
hand, with interleaved practice, performance in the short term won't sound
very good, but it benefits the learning of skills that will be long-term.

The CI effect, or interleaving, was first noted in word-pair learning during
the 1960s.[8] It became a major focus of research on motor skills, including a
wide range of sports such as badminton, golf, snowboarding, and tennis.[9] In
addition to being more effective than repetition for learning motor skills, it
has also been found to be good for cognitive tasks, improving math learning,
the study of physiology, and foreign-language learning.[10] In every case, improvement for the subjects using interleaved practice in these studies was
substantially higher than for those using repetition.

Although research on interleaved study in music is limited, a 2016 study
showed greater improvement in a group of advanced clarinetists after
interleaved practice than after blocked practice.[11] Yet, even though most
participants in the study found the interleaved schedule to be more useful,
they still preferred the blocked schedule. That feeling of fluency that comes
from repetition has a powerful impact on how we practice, even when we
have evidence that repetition doesn't lead to long-term retention of learning.

There are two hypotheses about why the contextual interference effect, or
interleaving, works. The first suggests that the two or three different tasks
being practiced together are both in working memory. Working memory
must then compare and contrast, which leads to stronger and more distinc-
tive encoding for each task.[12] The second hypothesis suggests a very different
scenario: switching from one task to another prompts the forgetting of the
first task's action plan. When you return to it, you must reconstruct it, and
that leads to stronger encoding in the brain.[13] There is also speculation that
both hypotheses may work in some kind of combination. Joseph LeDoux
puts it another way: interleaved learning prevents new information from interfering with old memories.[14] With interleaved learning, the representation
of the new learning is built up slowly over time and repetitions, adding to the
knowledge base rather than interfering with previous memories. Whatever
the brain process is, interleaving works.

Use of all the senses

The idea that we learn best when information is presented in one's preferred learning style has been around for a long time. Learning style theory may be presented in terms of sensory information (visual, auditory, or kines-thetic), or it may refer to mental activity (analytical, reflective, experiential, imagining), or dozens of other classifications. However, there have been multiple studies since the early 2000s that demonstrate there is no evidence to support the efficacy of any kind of learning style theory.[15] It is true that most people have a preferred way of taking in information. Many musicians learn by listening, using their auditory sense. But some are more tactile and learn by how the music feels as they play the instrument. One may absorb information more quickly through one particular sense, but evidence shows that the information is also forgotten more quickly.

When I was in college I learned by ear, and it wouldn't take too many times of playing through a piece before it was "memorized." I would sit in my dorm room and play through recitals in my mind before I played them on-stage. Initially, I had no problems, nor had I had memory problems in high school. But a friend said, "One day, you are going to crash and burn onstage because you just don't memorize right." I didn't know what he was talking about because I had never experienced any difficulties. But because he had planted the thought, I began to stress about something happening, and sure enough, it did. Slips in memory began to occur, and I discovered that my rapid learning of the music left some holes. I wasn't hearing the music in my mind as completely as I had thought. Yes, I was hearing most of a musical score, but I was sometimes a bit fuzzy on some of the internal voices. I had been fortunate for most of my life, up to mid-college years, that I hadn't had problems onstage. But that was sheer luck. As soon as I began to stress about my memory, instances would come up in performance when I wasn't sure about a particular chord or an internal line, or something would distract me, and I would flounder. Yes, you should hear the music in your mind, but you must hear *everything*, and you must have retrieval cues so you always know where you are.

It is important to also concentrate on the senses that we don't tend to use as "fallbacks." If you can't hear the music in your mind clearly, practice hearing one voice at a time, or a few bars at a time, until you can hear eve-rything. Nadia Boulanger, the great French music teacher, reportedly had students sing one voice of a Bach fugue while they played the other three (or more) from memory, alternating until they had sung all four voices. If you have trouble visualizing, practice writing out a small portion of the score from memory. See if you can visualize how complex chords look on the staff. Pianist Rebecca Shockley has written a wonderful book, *Mapping Music: For Faster Learning and Secure Memory*, that demonstrates how to diagram the main features of a piece and use it as a visual aid, or map, for learning music. It's a helpful tool for younger children, but also a good aid for experienced musicians.[16]

A voice professor colleague recounted an experience with a student who had great difficulty memorizing a particular aria. My colleague asked her to draw the music and then posted her drawings near the ceiling around the studio. Once the student had those visual prompts in her mind, she had no further difficulty with memory. It was another way of encoding information about the piece in her brain. Recent research has shown that drawing pictures or images of information that needs to be remembered is a more reliable strategy than mere repetition.[17] Opera singers say that thinking about the staging sometimes helps them remember the words—another visual prompt. The more ways in which information is encoded, the better the recall.

The "feel" of a piece in the muscles—embodied cognition—also needs to be encoded. The end of the second movement of the Schumann *Fantasie in C major*, Op. 17, is notoriously difficult for pianists (see Figure 6.1). The movement is marked "energetic throughout," with the last section marked "much more agitated." In the last twenty-eight bars, both hands play wide leaps in

Figure 6.1 Excerpt from Schumann *Fantasie in C major*, Op. 17, second movement

a dotted rhythm. A pianist has neither time nor capability of seeing where both hands are at the same time. This section requires physical stamina and unerring marksmanship. One well-known pianist spoke of a performance in which, after missing the first leap, he went on to miss most of the nearly 100 leaps to the end of the movement. That's not an experience one would want to repeat, but when stressed, muscles can tense up, and that changes one's spatial perception. When I first learned that movement, a fellow pianist suggested that I either practice with my eyes closed or with the lights off so I would learn to use my kinesthetic sense—to feel the distance my arms had to move to be accurate. The kinesthetic sense is greatly developed in blind pianists because they must be able to "feel" distances in their bodies—embodied cognition. But we all need to make developing our kinesthetic sense a part of our practice routine.

String players need spatial memory and a good kinesthetic memory when they shift positions. A violinist going from a B in first position on the A string to a D in third position must know both what it sounds like and what it feels like. The hand is shifted as a unit, sliding lightly along the string, but there is no fret on the fingerboard to indicate where that D in third position is. The hand must remember the amount of space between positions and what it feels like to move to that position. A timpanist cannot be looking at all four timpani at once and must know/feel the amount of space between them, nor can a mallet percussionist see the entire range of the xylophone or marimba at once. It must be felt in her body. A trombone has seven positions, and they must be learned by feel. Developing spatial awareness and a strong kinesthetic memory is important for many instrumentalists. *[On the companion website, there is a link at item 6.1 to an article about embodied practicing by flutist Lea Pearson ⊙.]*

Extremely slow practice

Many of us were told and continue to believe that to perform very fast passages, we must practice slowly and build up speed. Slow practice is good, but not for building up speed. Influential piano pedagogue Abby Whiteside declared that "Slow practice can establish habits that are completely unrelated to the coordination demanded for speed."[18] Whiteside may have known the work of Karl Lashley, a psychologist specializing in learning and memory, who observed that fast movements don't allow for planning each component, and that to play fast sequences, there must be a single motor plan encompassing the entire passage.[19] That is confirmed by neuroscientist Eckart Altenmüller, director of the Institute of Music Physiology and Musicians' Medicine in Hannover, Germany. Altenmüller suggests that different brain areas are responsible for slow and fast movements. Slow movements are "under steady sensory control," while fast movements "have to be performed without online sensory feedback," meaning that no one can monitor the notes as fast as he can play them.[20]

A slow movement pattern is established very consciously, movement by movement, under control of the frontal cortex. As soon as the movements begin to be automatic, they are transferred to the basal ganglia, and Altenmüller refers to the storage of these movements in the basal ganglia as a kind of zip file. The movements are packaged together, the brain grouping small movements into larger sequences; when played at a fast tempo, there is no time to think of each individual movement. The transfer from slow movement to fast movement is not continuous, just as the transition from walking to running is not continuous. It involves a different movement pattern. Fast movements are organized differently in terms of patterns and in terms of gravity. So, practicing at a slow tempo may hamper playing the passage at a fast tempo. One must practice fast to play fast.[21]

Whiteside, Lashley, and Altenmüller were talking about motor skills. Although slow practice may not be beneficial for motor skills, it is extremely helpful in terms of solidifying declarative long-term memory. It's a kind of testing strategy. Can you play a piece you know well from memory at a quarter of the tempo? If a quarter note is M.M. 60, try playing it at sixteenth note equals M.M. 60. The extremely slow tempo forces one to think about what comes next. There is no way to play it automatically, no using muscle memory or the movement patterns used when playing fast. Every single note, chord, rhythm, articulation must be considered—something not possible at a fast tempo. Perhaps the rhythm has slight alterations from the first theme to the second. Or maybe a double stop is added in the second theme. If you play excruciatingly slowly, you must think about those differences. If you can't remember, look it up, take note of the information, and the neural pathways become more securely consolidated.

Change the context

Psychologists have known for some time that people have better memory recall when tested in the same environment in which they studied. If you have studied with music playing in the background, you will test better if that same music is playing at the time of the test. People will test better in the same room. Music, lighting, paint color on the walls, the kind of chair one is sitting on, all provide contextual cues, both conscious and unconscious.

Many studies have tested this theory and most of them have to do with recalling words under different conditions. Recalling words isn't the same as recalling music, but both are about recall of information. A 1985 study found that students who listened to music while studying a list of words were able to recall more words when they were tested while listening to the same music.[22] Another study demonstrated that students who smoked either real marijuana or a placebo joint (looked and smelled real but no drug) scored significantly higher on a test of word recall when their brain was in the same state during testing as during study. In other words, whether they smoked a real joint or a placebo, they scored higher on the test when they smoked the same thing at test time.[23]

Another study from the 1970s has particular relevance for musicians, again involving word lists. Two groups of students studied a list of forty words in two ten-minute sessions that were a few hours apart. The first study session for all the participants was in a small, cluttered, windowless basement room (like most practice rooms). During the second study session, half of the participants studied in a comfortable room with windows overlooking a courtyard. Three hours later, they were all asked to write down as many words in ten minutes as they could remember from their previous study sessions. This time they were all in a classroom that was quite different from either of the study session rooms.[24]

The second group, that had studied in two different environments, recalled 40 percent more words than the group that had studied in a single room. This is significant. Researchers don't know exactly why, but the brain may encode some words in one environment and others in a different environment, or more contextual cues may be added by studying in more than one place.

Musicians rarely perform in the same room as where they practice. They practice in a practice room, or perhaps a living room or home studio. Ensembles frequently practice in rehearsal rooms. In their usual practice space, musicians tend to concentrate on technical issues, memory, and interpretation, and all of that is encoded in the context of that space. When they enter a hall or performance space, there is suddenly a different environment with different acoustics and lighting; if you are a pianist, a different instrument. The musicians must learn how they sound in that space and how to project in a larger space. Singers are suddenly faced with totally different visual cues. Soprano Lynn Eustis has pointed out, "there's no way to forget that you are singing at Carnegie Hall when you face that audience and open your mouth."[25] If the acoustics are lively, a pianist or string player must consider pedaling or bowing differently. Different issues with the music must be confronted in the context of a different space and all of that must be added to the information already known about the music one is performing. More contextual cues are added to the neural circuitry that already exists.

Of course, this isn't a conscious process; contextual information is always added when information is encoded in the brain. That's why smells sometimes trigger memories because the smell has been encoded along with the memory. So obviously the more venues one has practiced in (and, for pianists, the more instruments played), the more contextual cues there are—which aids recall. Mood is one of those contextual cues. Practice usually happens when one is in a calm mood, stress kicks in at performance time. Therefore, it is a good idea to simulate "performance" events that will raise stress level. Self-recording always adds a layer of stress. Play or sing for friends but treat it as an actual performance. Perform in the hall if possible. Even with no one there, set the lighting as it will be for the performance and walk on and off stage to create as much of the performance mood as possible. Bow to the imaginary audience. All of those added contextual cues help at the time of the actual performance.

Imagery

Imagery is such a powerful tool for learning that it will be discussed separately in Chapter 8.

And finally, practice meets sleep

> It is a common experience that a problem difficult at night is resolved in the morning after the committee of sleep has worked on it.
>
> —author John Steinbeck[26]

We tend to think of learning music as being synonymous with physical activity—that to learn a piece of music, you must physically practice it. But learning and memory formation continue long after physical practice ends, and we don't pay attention to this part because we aren't aware it is happening. "Let me sleep on it" usually refers to delaying a decision and gaining time to mull it over. But in the case of learning and memory, adequate sleep is necessary (1) to prepare the brain for encoding new material; (2) to consolidate memory more strongly; and (3) to ensure access to memory when under stress (as is common during a performance).

As long ago as the 1920s, researchers found that memory retention was better after a night of sleep than after an equivalent amount of time awake. They thought this was because the brain wasn't receiving any sensory input during sleep that would interfere with what had been learned earlier. That the brain might be actively doing something during sleep wasn't even considered.[27] But with the discovery in the 1950s of rapid eye movement (REM) sleep and non-rapid eye movement (NREM) sleep, researchers discovered that sleep isn't a single state. While we sleep, our body cycles through several stages of both REM sleep, which is when we dream, and NREM sleep. NREM ranges from the lightest stage of sleep, N1, to the deepest stage, N4. N3 and N4 are known as deep sleep or slow wave sleep (SWS) and characterized by delta waves, the slowest brain waves. Over the past twenty-five years, researchers have learned a great deal about the role of specific stages of sleep and how each may contribute in a unique way to memory processing.

Matthew Walker, leading researcher on sleep and author of *Why We Sleep: Unlocking the Power of Sleep and Dreams*, comments that given the number of stages in memory, the multiple kinds of memory, and the several stages of sleep, "one is faced with a truly staggering number of possible ways that sleep might affect memory consolidation."[28] Nonetheless, there are a few findings concerning the role of sleep in facilitating encoding and consolidation of memory that are of particular interest to musicians. According to Walker, consolidation begins while we are practicing and continues from a period of a few minutes up to six hours after practice while one is awake. This period is called consolidation-based stabilization. The memory is stabilized and maintained at the same level as when practice ended. It does not improve. On the other hand, that memory is enhanced after a night of sleep—additional learning takes place without any additional practice, and Walker calls this period consolidation-based enhancement.[29]

Sleep and consolidation of memory

Sleep *after* learning is important for memory consolidation, for making memory stronger and more stable. There is substantial evidence to indicate that both SWS and REM sleep contribute to the consolidation of declarative memory, with differences depending on the difficulty of the task and whether the memory is of facts or events.[30] Consolidation of procedural memory is correlated with the amount of stage 2 NREM sleep, particularly in the last quarter of the night. Several studies have shown that motor sequence tasks improve in speed and accuracy after a night of sleep.[31] One of the most interesting studies of the effect of sleep on motor skills was done in 2013 by Sarah Allen, then at the University of Texas, now at Southern Methodist University. Allen used sixty undergraduate and graduate music majors divided into four groups in her study. They were majoring in other instruments but had rudimentary piano skills. They all learned one or both of two melodies on the piano in an evening practice session where they were monitored for speed and accuracy. They then went home to sleep.[32]

The students who had learned one melody, melody A, showed over 11 percent improvement in speed and accuracy when tested the next morning—without any additional practice. Those who learned two melodies, A and B, showed no improvement in either one. Learning a second melody seemed to cancel out the gains in the first. Group 3 had learned both melodies but reviewed the first melody before going home to sleep, and they showed about the same amount of improvement as the first group—11 percent. The last group, who learned A at night, B in the morning, and were then tested, showed no improvement. In a much earlier study from 1994, subjects who were deprived of stage 2 NREM sleep showed pronounced deficits in motor performance.[33] The value of sleep for *consolidating* memory of motor skills seems clear, but sleep is also important for *encoding* memory.

Sleep and encoding of memory

Sleep is important *prior* to learning for the brain to be prepared to properly encode declarative information. Sleep deprivation impairs encoding of memory. Several studies by Sean Drummond and colleagues found that sleep deprivation prior to verbal learning caused changes in how the brain encodes information, with the medial temporal lobe (hippocampus) not engaging normally and prefrontal areas of the brain (short-term and working memory) and parietal lobes (episodic memory) overcompensating.[34] If areas of the brain involved in encoding memory are not functioning normally, encoding will not be as efficient or secure. It's somewhat like painting a wall with your non-dominant hand. You may get the job done, but there will likely be some missed spots or messy places. The brain needs to be functioning at full capacity when trying to encode new musical material. If material isn't encoded, it can't be consolidated, and you won't be able to retrieve it at performance time.

Sleep and anxiety

Anxiety and stress are often associated with performance, but consistent deep sleep reduces anxiety. A study at the University of California, Berkeley, showed that a sleepless night can raise anxiety levels by up to 30 percent, and that is multiplied after several sleepless nights.[35] The part of the sleep cycle that is important for "resetting" the brain and reducing anxiety is slow-wave sleep, the deepest part of the non-REM cycle.

Nearly all stages of sleep are important for the encoding and consolidation of both declarative and procedural memory. Swedish researchers studied both procedural and declarative memory under sleep deprivation. They found that with only four hours of sleep, both kinds of memory were fine. Theoretically, performing on four hours of sleep should be okay. But that was without stress. When the participants were subjected to stress as well as sleep deprivation, procedural memory remained intact, but there was significant impairment of declarative memory.[36] Add stress from lack of sleep to the stress of performing, and chances are that performance will suffer. You won't forget how to play your instrument, but there is a very good chance you will forget the music. Few of us perform without some degree of stress, so it's a good idea to counter that with plenty of sleep. Even a sixty-to-ninety-minute nap significantly enhances procedural memory consolidation.[37]

Forming lasting memory for a piece of music depends on brain plasticity, on lasting functional and/or structural changes in response to practice, and on neural networks becoming stronger as the synapses become stronger. These changes occur not only when actively practicing, but also while asleep. Sleep could well be the most important part of the memory process.

How much practice is necessary?

The idea that it takes 10,000 hours or ten years of practice to achieve expert performance in any field has been floating around for almost thirty years. It was popularized by Malcolm Gladwell in his 2008 book *Outliers: The Story of Success*, but the idea wasn't Gladwell's, nor did any researcher ever claim that "ten thousand hours is the magic number of greatness," as Gladwell claimed.[38] The 10,000 hours idea came from psychologist K. Anders Ericsson in a 1993 study of student and professional violinists in Germany.[39] Music professors at the Music Academy of West Berlin had nominated fourteen violin students with the potential to become international soloists to be participants in the study. Professional violinists from the Berlin Philharmonic and the Radio Symphony Orchestra also participated. They were all interviewed about the age at which they began violin study, the number of hours of deliberate practice per week, the number of hours spent in other musical activities, participation in competitions, and more.

Ericsson found that by the age of twenty, both professionals and students in the professional track had accumulated an average of 10,000 hours of deliberate practice. This compared to 7,500 hours for students who had been identified as "good" rather than professional, and 5,000 hours for student violinists who were in the music education department and not intending to be performers. Gladwell picked up on the 10,000 hours idea but didn't include the idea of deliberate practice.

There is no question that becoming a top-level performer is dependent on thousands of hours of practice—there are no shortcuts. Even for someone without professional aspirations, learning to play an instrument or sing well requires a lot of practice. There is also no question that the kind of practice matters—one must be attentive to details, focused, analytical, and willing to make challenging choices about how to spend one's practice time to achieve long-term gain. At some point in my student life, I played in a master class taught by renowned pianist György Sebők. After the class, Sebők sat and chatted with those of us who had performed, and the subject of the conversation eventually turned to practice. He commented that he didn't think it was possible to focus and pay the kind of attention needed for quality practice for more than three hours a day. He went on to say that every minute of those three hours needed to be focused on the music, on details, on analysis, on how to communicate the essence of the work—deliberate practice, not simple repetition. That kind of practice may be hard work, but it doesn't need to mean devoid of pleasure.

Find the joy

I was extremely fortunate to study with pianist Ann Schein while I was working on my doctorate at the Peabody Institute. Lessons with her were always an adventure in discovery, whether the current work being studied was Scriabin, Mozart, or Rachmaninoff. When I brought the Ligeti Études to a lesson, she was fascinated by these pieces she had never heard, and she discovered many interesting details on first hearing. Performances by this much-loved pianist have been described as intimate, powerful, elegant—and radiating joy all the way to the back row. In fact, the concept of joy infuses her whole approach to the piano, including practicing. Schein studied with Arthur Rubinstein and still speaks about "getting a lesson from Arthur" when she listens to one of his Chopin recordings. When learning that she had less than a month to learn the Mozart Concerto in E-flat Major, K. 449, she spoke about accepting the challenge "with great joy." She went on to say that she discovered "a revelation a measure," that she felt as though Mozart was teaching her something in every measure, some small or new detail that was instructive and fascinating.[40] Finding that kind of pleasure and magic in music you are studying makes it less about "deliberate practice," and more about stimulating discovery, about what there is in a piece that is compelling at this particular moment in life, about absorbing every marvelous detail of the music in order to communicate its unique qualities to others. In Schein's philosophy, learning a new piece of music becomes a joyful challenge.

Key Concepts

- Forgetting is an important part of the learning process—it creates space to add more facts and details to the memory.
- Storage strength refers to how well a piece of music is encoded; retrieval strength refers to how accessible that information is.
- The brain has unlimited storage capacity, but the theory of disuse says people forget what they don't use. The memory isn't lost, just currently inaccessible.

- The harder the work to retrieve a memory, the stronger it becomes. Therefore, practice strategies that require work, that present *desirable difficulties*, are needed.
- The two best strategies for learning and memory are practice testing and distributed practice.
- Other good practice strategies are interleaved practice, incorporating information from all the senses, extremely slow practice, and varying the context in which one practices.
- Sleep may be the most important part of the memory process.

7

Neuroplasticity—Awe-Inspiring to Debilitating and Back Again

I'm not a deaf musician. I'm a musician who happens to be deaf.
—percussionist Evelyn Glennie

Passion drives neuroplasticity. The more we love making music, the more motivated we are to learn, and the more readily neuroplasticity happens in our brains. Neuroplasticity makes it possible to learn an instrument, re-fine our technical skills, learn and memorize music, and if we practice long enough, become skilled performers. But neuroplasticity, driven by passion, can do much more. Neuroplasticity makes it possible for a blind pianist to develop an international career, including playing with orchestras, even though he can't see the conductor, the keyboard, or read a printed score. Neuroplasticity makes it possible for a percussionist to become the most famous drummer in the world, even though she has been profoundly deaf since the age of twelve. And for an award-winning drummer who suffered a traumatic brain injury and couldn't remember how to play the drums, neuroplasticity made recovery and a return to performance possible. These are a few of the awe-inspiring positive stories of neuroplasticity.

Yet neuroplasticity has a potential dark side for musicians—focal dys-tonia, a condition in which neuroplasticity becomes maladaptive, creating involuntary movements that make it impossible to continue playing. One to two percent of all musicians develop focal dystonia, and although there are various treatments and medications, doctors usually say there is no cure, and one will be unable to continue performing. Some musicians suffering from dystonia reinvent themselves to make music in other ways. A few leave the music world entirely. But some who develop the condition say, "Not so fast. Music is my life, and I am determined to find a way through this." They

The Musical Brain. Lois Svard, Oxford University Press. © Oxford University Press 2023.
DOI: 10.1093/oso/9780197584170.003.0007

use their passion for making music—and their patience, to "relearn" how to play, bypassing the maladaptive brain pathways and using the power of neuroplasticity to create new neural networks, thus being able to return to their careers. Neuroplasticity plus passion for making music is a powerful combination.

Steve Mitchell—a drummer and traumatic brain injury

Steve Mitchell was a popular West Coast studio drummer for nearly three decades. He had arrived in San Francisco in 1967 with the Skyliners, a band from Pittsburgh. His skill as a drummer quickly came to the attention of the television industry, and he played for programs such as *Garfield the Cat*, *Sesame Street*, *Nova*, the last twelve Charlie Brown TV specials, thousands of commercials, and was even a featured soloist for eight years with the Joffrey Ballet. Steve was recognized by the San Francisco chapter of NARAS (National Academy of Recording Arts and Sciences, now the Recording Academy and the Grammy Awards) as the "Best West Coast Studio Musician in 1975." (The story that follows is based on my conversations with Steve Mitchell between 2012 and 2019.)

Steve had a formidable technique, but there was also a spiritual side to his drumming. He believed that music stemmed from the human heartbeat and was therefore necessary to life. He often remarked that the primary reason he played was because drumming fed his soul, and he saw drumming as a way to put people in touch with a deeply spiritual experience. His life was all about music.

Steve grew up in a Quaker family on a dairy farm in North Central Pennsylvania. At some point in the mid-1990s, Steve moved back to that area, he thought temporarily. But one day, an accident involving a rather large woodpile in the forested area where he was living left him with a shattered right shoulder and a traumatic brain injury. In an instant, the unthinkable had happened. Steve spent the next nine weeks hospitalized in a coma, and when he slowly awoke, he could no longer walk, feed himself, talk, or, to his alarm, play the drums. He spent several months in rehabilitation, beginning the slow process of relearning how to talk, eat, walk—all the basic activities required for independent living. Still, what concerned him the most was making music, because his ability to do that seemed to have disappeared.

I met Steve and we became good friends a few years after that accident. He was a bear of a man with an infectious laugh, large beard, and hair that flew in every direction, but when he spoke about the accident, it was with a soft, reflective voice. He spoke about how he had not known whether he would ever be able to play again, about his muscles not remembering how to make the basic motions involved in playing a drum, about hearing music in his mind that he wanted to play but not being able to connect those sounds to the physical motions necessary to play them. Making music had been Steve's entire life, and he was devastated by the possibility he may never be able to play the drums again.

Steve couldn't remember the motions necessary to play the drums because, as we have seen, practicing and making music doesn't happen in the muscles. Practice happens in the brain, whether you are an athlete or a musician. Unlike athletes, however, the motor movements involved in musical practice must also be connected to sound, and the sound-motor connection in Steve's brain was no longer functioning. The brain must be able to connect a sound with the motion necessary to make it, or one won't be able to play the drums, the piano, or any other instrument.

Although the injuries were to his right shoulder and right side of his head, Steve didn't have the use of either hand. The right hemisphere controls the left side of the body, so the damage to the right side of his brain affected the sound-motor connection on the left side. The shattered right shoulder also meant that motor movements and sound were no longer coordinating on the right side of his body, controlled by the left side of his brain. Steve's shattered shoulder could be repaired, but that wasn't going to restore the sound-motor connection in his brain. Was repairing that connection even possible?

By the mid-1990s, many researchers had shown that some brain areas changed in response to practice (see Chapter 4). But the connection between the brain's ability to change and applying that knowledge in a clinical setting had not yet been made. In fact, as late as 2010, two researchers from the McKnight Brain Institute published an article that pointed out that because of medical advances, increasing numbers of people were surviving brain injury, but neurorehabilitation hadn't kept up.[1] If people in the rehabilitation field were still writing in 2010 about the problem of connecting neuroscience research to clinical settings, it is all the more remarkable that in the late 1990s, a physical therapist in a small-town rehab center in rural Pennsylvania devised a treatment that proved to be exactly what was needed to restore Steve's playing ability.

Steve's recovery

Steve's imaginative physical therapist not only asked a lot of questions about drumming to understand what was involved, but also watched drumming videos to see for herself what range of motion Steve would need depending on the kind of sound he wanted to make. She created exercises that encouraged his brain to use the same neurobiological processes it used when he first learned to play the drums, matching specific sounds to specific movements for every drum and cymbal in his kit. At one point, she even asked, "don't you have a cymbal way over there?" pointing to his far right, and then designed an exercise that would help him move his arm in that direction and connect with the sound of that cymbal.

And Steve began to practice. At first, he could physically manage only five minutes a day, but little by little, he added time. Just as when he had first learned to play drums as a child, he was again having to match specific sounds with the movements to make those sounds and match sounds to different touches on the drum made by his fingers, hands, or sticks. But when children learn to play an instrument, they don't have sophisticated musical models in their minds to guide them. They are building from the ground up—both the ability to hear music in their minds and the ability to match sound to movement. Steve had decades of sound experience in his mind. He knew exactly what kind of sound he wanted. His brain just needed to create new neural pathways to connect the sounds he heard in his mind to the movement necessary to create those sounds, to reconnect sound and motor programs. The damaged pathways between the auditory and motor regions of his brain could not be reconstructed or reused, so he was creating new pathways through the exercises.

Since Steve had always thought of drumming in spiritual terms, his exercises became a form of meditation to connect him with that spiritual element. It took about a year for him to feel comfortable playing again in public, and then he began to add performance gigs, a few at first and then increasingly more and longer ones. He had created, and continued to create, new pathways connecting the auditory and motor areas of his brain, bypassing the damaged areas—neuroplasticity at work.

In the process of relearning to play and rebuilding his formidable technique, Steve looked at the essence of his musicality and realized that "less is more." He thought about the silence he experienced in Quaker meetings as a child and realized during his year of meditative practicing that he could do more with his playing by not making it all about technique. With less emphasis on technique and more on feeling, his playing became more emotionally communicative. His focus changed. Music in his life became about building community. Steve is seen in Figure 7.1 several years after the accident.

Steve had received a bachelor's degree in music education from Duquesne University prior to moving to San Francisco. Since he now had no interest in returning to the frantic life of a studio musician, he began teaching again, spending a year at a Friends school teaching the students to communicate through drumming. He began teaching private students, some of whom went on to music schools and conservatories. He played with professional groups, but also with local groups, always with the emphasis on emotional communication, not technique. Building community was important to him,

Figure 7.1 Steve Mitchell, drummer
Credit: Jeff Solomon Photo

and he continued to lead drum circles for that purpose until shortly before his death in 2019.

Few musicians suffer the kind of brain damage that Steve did, but his recovery shows the power of neuroplasticity to reshape brains, as well as lives. Learn an instrument, change your brain. This happens not only when originally learning but also when forced by circumstances to relearn.

Cross-modal neuroplasticity

Steve had to learn to reconnect motor and sound programs, but his senses had remained intact. He could still see his drums and hear music live or in his imagination. But what happens in the blind or deaf—if the optic or auditory nerve is damaged or non-functional and cannot relay sensory information to the visual or auditory cortex for processing? Hearing- or vision-impaired individuals can still become extraordinary musicians because brain neuroplasticity makes it possible for one sense to substitute for another. This is *cross-modal plasticity* in which hearing or touch serve as a blind person's eyes; touch or vibrational sensations serve as a deaf person's ears. Cross-modal neuroplasticity is a kind of functional neuroplasticity (see Chapter 4) because a sensory area changes function, repurposing itself to process signals from a different sensory input—the visual cortex processing touch signals, for example.

We identify certain brain areas with specific functions. The theory of *localization* of brain function began with Pierre Paul Broca's discovery in 1861 of the area of the brain responsible for language. By the beginning of the twentieth century, many other brain areas had been identified: the motor cortex, somatosensory cortex, visual cortex, and auditory cortex, among others. Wilder Penfield's experiments that mapped the motor cortex and the somatosensory cortex (the homunculi) seemed to confirm that different areas of the brain were responsible for different functions, such as vision or hand movement (see Chapter 4). In localization theory, for example, if the visual cortex was unable to receive sensory input from the eyes, it would cease to function.

But a neuroscientist named Paul Bach-y-Rita had a different idea. He believed that since all sensory information enters the brain as patterns via nerve fibers, any of the specialized areas should be able to process neural signals coming from any of our sense organs—eyes, ears, nose, tongue, and skin. Those patterns are, in some way, all equivalent. In the late 1960s, he introduced the idea of sensory substitution, stimulating one sense to take the place of another. If one sensory area in the brain is damaged, such as the auditory cortex, another could substitute for it and take its place.

Over a period of years, he built a series of devices that would allow blind people to see, or those with damaged vestibular systems to regain their balance and live more normal lives, one sense substituting for another—sensory substitution. These devices, which collect data from the outside world and transform it into electrical signals that the brain can interpret, are known as sensory substitution devices, or SSDs.[2]

The basis of sensory substitution is a kind of neuroplasticity called *cross-modal reassignment* in which the brain reorganizes itself so that a sensory area that has been deprived of its main inputs receives inputs from another sensory source. The SSDs that Bach-y-Rita and others built provide an external assist for neuroplasticity. But individuals who are congenitally blind or deaf, or who lose those abilities at an early age, often develop cross-modal neuroplasticity to an extraordinary degree without assistive devices, learning to use touch to "hear," for example, or enhanced hearing or spatial navigation to "see."

In previous decades, it was common to speak of someone who lacked sight or hearing as disabled, even impoverished. It was thought that being without one sense would lead to overall cognitive impairment. Although a blind or deaf individual may not experience the world in quite the same way as a sighted or hearing person, it has been clear for some time that their brains make amazing accommodations allowing them to function well without sight or hearing. Researchers have found these changes involve enhanced processing in the remaining sensory areas of the brain. Additionally, the areas that would normally process the missing sense repurpose themselves to process a different sense—cross-modal neuroplasticity. These neuroplastic changes sometimes translate into behavioral skills that are equal to or even superior to those of individuals who haven't lost those senses.[3]

Visual loss

There have been many famous blind jazz, blues, and popular musicians. George Shearing, jazz pianist; Ray Charles, soul music pioneer; Lennie Tristano, jazz pianist and composer; and Art Tatum, perhaps the greatest jazz pianist of all time—were all marvelous musicians and had outstanding performing careers. Stevie Wonder, who became blind shortly after birth, is considered a genius who has influenced the genres of rhythm and blues, soul, and jazz. Marcus Roberts, who lost his sight by the age of five, is a highly acclaimed pianist and composer, his music influenced by jazz greats while blending in elements of classical traditions. Although the lack of sight in all these musicians is occasionally talked about, it seems to be casually accepted in the world of jazz or blues. Blindness has rarely, if ever, been much of a topic in reviews of their performances. Critics comment on the music and how they play, not that they are blind. The classical world is another story.

The first blind classical musician to medal in a major international competition was Japanese pianist Nobuyuki Tsujii, seen in Figure 7.2, who shared the gold medal with Chinese pianist Haochen Zhang at the 2009 Van Cliburn International Piano Competition. The general response in the classical music world was "how can a blind pianist play so flawlessly and beautifully?" Some jurors at the Cliburn Competition spoke of being so moved by Tsujii's performance, they left the room in tears. Critics who have reviewed his concerts and conductors with whom he has played have spoken about his expressive musicianship, his unfailing assurance, the sincerity in his playing, and how

Figure 7.2 Nobuyuki Tsujii, pianist
Credit: Stephanie Black for WQXR and The Greene Space

miraculous it is that he plays so divinely despite being blind. Reviews always speak of his blindness, which rarely happens with blind jazz musicians. How is Tsujii able to learn repertoire without being able to see the score, play with an orchestra when he can't see the conductor, and make the huge leaps found in some piano works without being able to see the keyboard? [*A link to Tsujii's performance at the Van Cliburn Competition of the first four Etudes, Op. 10 by Chopin can be found on the companion website at item 7.1 ⊚.*]

Individuals who are blind have been shown to have a finer sense of touch, to have superior and more efficient auditory-pitch discrimination skills, better discrimination at localizing sound in space (how near or far away is the sound), and are superior in spatial navigation. Finger areas in the somatosensory and motor cortices are expanded and reorganized in blind Braille readers due to increased use, just as those finger areas have been found to be expanded in pianists and violinists. The auditory cortex is also expanded in blind individuals. Existing sensory areas take on an increased role in the blind and become expanded and reorganized with increased use.[4]

But the visual cortex has also been found to process hearing and touch signals, and this is cross-modal neuroplasticity. In 2008 at the Center for Noninvasive Brain Stimulation at Harvard, researchers scanned the brain of a congenitally blind painter to try to find out more about neuroplasticity.[5] The lead researcher was Alvaro Pascual-Leone, the same neuroscientist who demonstrated in the early 1990s that the adult brain could change as the result of practicing a five-finger exercise on a keyboard (see Chapter 4). Now Pascual-Leone was looking at the brain of Eşref Armağan, a Turkish painter who has been blind since birth. Armağan paints with his fingers. He paints extremely realistic landscapes and portraits that show perspective, color, and detail, and his paintings hang in galleries and museums around the world. One of the researchers in the study had already demonstrated that in blind individuals the sense of touch recruited visual areas to produce a representation of an object in the brain.[6] [*Examples of Armağan's paintings can be seen on the companion website at item 7.2 ⊚.*]

Armağan's brain was scanned while he lay in an fMRI scanner drawing small objects that he touched and handled before beginning to draw. To the surprise of the researchers, his visual cortex lit up as though he were sighted in exactly the same way as if it had been receiving signals from the optic nerve, though his was non-functional. Touch was triggering his visual cortex—cross-modal neuroplasticity. In blind individuals who have a damaged optic nerve or some disorder of the eye itself, there are no visual signals for the visual cortex to respond to, so it repurposes itself to respond to touch and auditory input instead. Researchers have found that when blind subjects are reading Braille, their brains are showing activity in the same part of the visual cortex as sighted people who are reading, so the visual cortex is responding to touch.[7]

Comments made by Tsujii in many interviews and documentaries make it clear that his sense of hearing is heightened. No doubt the auditory cortex itself has taken on an enhanced role. It is also likely that, just as a blind painter's visual cortex would repurpose itself to respond to touch, a blind pianist's visual cortex would repurpose itself to respond to hearing. Tsujii learned Braille notation at the age of six, but quickly discovered that little repertoire was available in Braille, and it was a slow way to learn music. He could learn faster by ear.[8] Now when he learns a new piece, a team of assistants records the score in small sections of a few bars each, each hand separately, with verbal comments about the composer's markings and instructions. Tsujii learns the piece by listening to these recordings. He calls these recordings "music sheets for ears."[9] He learns each hand separately, then together, adding his own interpretation. The method seems daunting, but Tsujii does it easily.

In a 2017 interview with the Australian Broadcasting Corporation after performing the Chopin Second Concerto with the Sydney Symphony Orchestra, Tsujii was asked how he stays in time with an orchestra when he can't see the conductor. His reply, "By listening to the conductor's breath and also sensing what's happening around me."[10] Being able to hear the conductor breathing from several feet away in the middle of a concerto performance reflects a very heightened sense of hearing, as do Tsujii's comments that he "accounts for timing changes based on how the sound echoes in the hall."[11]

For a blind pianist, spatial orientation is extremely important. He must be able to make large leaps and move over the keyboard quickly. Tsujii orients himself spatially at the piano by feeling the highest and lowest registers before he begins, and then plays with his torso relatively still so that his spatial orientation does not change.[12] He always knows where he is in relationship to the keyboard. While most pianists look before they leap on the keyboard, that's a concept that he doesn't seem to understand. "People always ask me about leaps, but they are not difficult. For me the piano is an extension of my own body, so I know exactly where everything is."[13] This is embodied cognition, knowing with the body. Tsujii not only feels the music with his body, but the piano has become an extension of his body as well.

Not unexpectedly, Tsujii also has a heightened sense of touch and feeling. The inspirational documentary *Touching the Sound* by Peter Rosen shows Nobu (Tsujii's popular nickname) touching flowers, snow, a lobster, sand, water.[14] He learns about his world through touch and sound. He "feels the vibrations from the applause" and says that he feels the energy level of the audience and he listens for their enjoyment.[15] Cross-modal neuroplasticity makes it possible for Nobu to play the piano technically at an exceptional level, but ultimately, it's about connecting with the audience, and that seems to be Nobu's greatest joy.

Hearing loss

With the deaf, as with the blind, other senses are heightened. Deaf individuals have been shown to have greater tactile sensitivity to vibrations, enhanced peripheral vision, and heightened sensitivity to emotional expression in others.[16] These reflect enhancements in the sensory cortices that are functioning, such as visual and sensory. And in cross-modal neuroplasticity, auditory cortical areas, even though they are not receiving information from the ear, process signals from Braille reading (touch), sign language (visual), and vibrations.[17]

In research with students at the National Technical Institute for the Deaf in Rochester, NY, Dean Shibata found that, when feeling vibrations, both hearing students and deaf students showed brain activity in the part of the brain that normally would process vibrations, the somatosensory cortex, but deaf students also showed brain activity in the auditory cortex that the hearing students did not. Musical productions are apparently an important part of the culture at the school. Audience members are given balloons to hold during the performance so they can "feel" the musical vibrations. Vibrations carry the same information as sound waves, so they substitute for the sound waves that would normally be transmitted to the brain via the inner ear.[18]

No one better demonstrates hearing-through-touch than percussionist Evelyn Glennie, seen in Figure 7.3. Deaf since the age of twelve, she has spent her professional life on a mission to teach the world to listen. Because she has learned to hear in a different way, by being extremely attentive to touching and feeling, she wants others to experience sound more fully as well. In her 2015 *Hearing Essay*, she says that "hearing is basically a specialized form of

Figure 7.3 Evelyn Glennie on Marimba in Harrogate
Credit: James Wilson/The Evelyn Glennie Collection

touch." She writes that sound is vibrating air that the ear converts to electrical signals that are sent to the brain and interpreted. But vibrations can do the same thing.[19] Glennie experiences every day what researchers have learned in labs about cross-modal neuroplasticity. *[A link to Glennie's "Hearing Essay" is found on the companion website at item 7.3 ⊚.]*

Glennie has had an extraordinary career filled with "firsts:" she is the first person in history to successfully create and sustain a full-time career as a solo percussionist; she has commissioned more than 200 new works for percussion; she is a double Grammy Award winner and BAFTA nominee; she has recorded more than forty CDs; she gives more than 100 performances a year plus master classes; she has composed music for film, television, and theater; and she led 1,000 drummers in the opening ceremonies of the 2012 Olympic Games in London.

And yet, even though she is living a career that for a hearing person would prompt a totally different set of questions, interviewers most frequently ask her some version of "How can you be a musician if you can't hear?" The answer, of course, is that she does hear. She wouldn't have been able to have the career she has had if she could not hear. She just hears in a different way due to cross-modal neuroplasticity.

Glennie grew up on a farm in northeast Scotland. She had what she describes in her autobiography *Good Vibrations* as a very ordinary childhood, playing outdoors, attending church where her mother played the organ, helping care for farm animals, going to school. Glennie began to play the piano at age seven and added the clarinet two years later. Meanwhile, she had begun to lose her hearing due to nerve deterioration, and by the age of twelve, she was profoundly deaf. She decided she wanted to learn percussion after being introduced to the xylophone at a school assembly. She wasn't thinking about becoming a professional percussionist—she just loved the instruments.[20]

She was extremely fortunate to have had a school music teacher, Ron Forbes, who encouraged her to expand how she heard sound. "He would say to me 'Evelyn, create the sound of the sun radiating on your face' and I wondered—how was I to do that? What he was really asking me to do was to express the feeling of sound."[21] In a 2003 TED talk titled "How to Truly Listen," Glennie demonstrates that listening involves much more than your ear.[22] She talks about her music teacher helping her learn to feel vibrations throughout her body by holding her hands against the wall as increasingly narrow intervals were played and feeling those vibrations in her body. That, in fact, is how Glennie hears, through her fingers, her arms, her face, her tummy. She plays barefoot, so she can feel the vibrations better through her feet. It is clear from her descriptions that her sense of touch and her feeling of vibrations have taken the place of her ear. In fact, she speaks of her body as a huge ear. *[A link to the blog post "I Have Seen and Touched the Sound" is found on the companion website at item 7.4; a link to Glennie's TED talk, "How to Truly Listen," is found at item 7.5 ⊚.]*

Since Glennie didn't become deaf until after she had learned to play the piano, she can remember the sounds of music from before she lost her hearing. That means that when she looks at a score, she can hear it all in her mind. She had already learned to speak before she lost her hearing, so she speaks with the inflections of a hearing person, unlike those deaf from birth, and she lip-reads. Her auditory cortex is activated by vibrations, just as blind painter Eşref Armağan's visual cortex is activated by touch, and blind pianist Nobuyuku Tsujii's visual cortex is triggered by both touch and auditory signals. She has written and spoken extensively about how she hears, and how everyone should open themselves to "feeling" sound, not just using their ears. Glennie says, "I know how music sounds by what I feel and see. I can sense musical sound throughout my whole body. I can identify different pitches in isolation according to which part of my body feels the vibrations and for how long."[23] *[See link on the companion website to Glennie's "Deaf, Sound and Music Questions," item 7.6 ⊕.]*

The documentary about blind pianist Nobuyuki Tsujii is titled *Touching the Sound.* Ironically, a documentary about Glennie is titled *Touch the Sound.*[24] The similar titles reflect the importance of touch to both musicians.

Focal dystonia

Neuroplasticity is the brain's great ability to improvise, to make the connections necessary for us to be able to learn to make music, to find new ways to make those connections when existing networks are damaged, and to repurpose a sensory cortex when one of our senses is missing. But that great improvisational ability of the brain sometimes goes awry. That's the case with focal dystonia.

Dystonia is a general term for a large group of movement disorders that are characterized by uncontrolled movements and involuntary muscle contractions. "Focal" means the dystonia affects a specific part of the body, in musicians, usually the hand or embouchure. It is called task-specific focal dystonia because it occurs only while a musician is playing his instrument and usually has no impact on any other activities. It affects between 1 and 2 percent of professional musicians and it can have a devastating impact on a musical career. Pianists, guitarists, and violinists develop hand dystonia, which causes loss of fine-motor control in the hand. It affects the right hand in pianists and guitarists, the left in violinists—the hand in each instrument that has the higher workload.[25] Brass and wind players are at double risk. They can develop hand dystonia, but they can also develop embouchure dystonia, which affects the muscles of the face, mouth, tongue, and jaw, preventing them from forming an embouchure. Singers can also develop focal dystonia which remains confined only to singing, not speaking, and occurs only during specific tasks. It has been studied far less than other musicians' dystonias because it is often attributed to problems of vocal technique.[26]

Hand dystonia begins with a feeling that one's fingers just aren't working quite right, a finger or two may feel weak, passage work gets a bit sloppy, there may be tightness in the arm, and eventually one or more fingers cramp and curl under, making it impossible to play. Embouchure dystonia often begins with a feeling of not being able to control one's embouchure, entrances aren't clean, pitches aren't centered, and eventually it can lead to jaw clenching, severe lip tremors, and loss of control of the tongue, again making it impossible to play. Because it is task-specific, occurring only when you are playing your instrument, and because it is painless, most musicians mistake the early signals and assume the problems must be from too little practice, leading them to practice more diligently—which makes the issues worse. In the process of trying to discipline recalcitrant fingers, many individuals overuse or overextend other fingers, change the position of hands or wrist, sometimes developing tendinitis or other conditions that cause pain and make the symptoms more severe. With embouchure problems, the player may press the mouthpiece harder into the lips, try to "shape" the embouchure, and end up creating more tension. Patterns of dystonic posture can be seen in Figure 7.4.

Although the conditions that came to be known as dystonia were described in the medical literature in the mid-1800s, the causes of focal dystonia remained elusive, with one branch of science believing it could be traced to changes in the nervous system and another, following in the footsteps of Sigmund Freud, believing emotional distress caused the physical symptoms

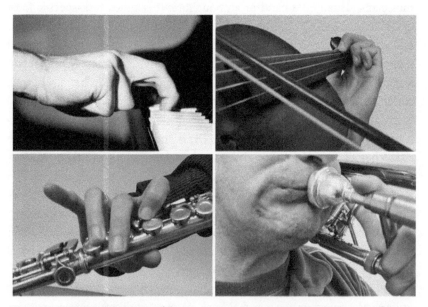

Figure 7.4 Typical patterns of dystonic posture in a pianist, a violinist, a flutist, and a trombone player

Credit: Photo used by permission of Eckart Altenmüller

of dystonia.[27] For much of the twentieth century, researchers and medical doctors didn't really understand the cause of, or the treatment for, patients who suffered from any of the multiple kinds of dystonia. It was often classified as a psychological rather than a neurological disorder.

Pianist Leon Fleisher had the misfortune to develop focal dystonia at a time when it was still thought by many to have psychological roots. On his way to becoming one of the most important pianists of his generation, Fleisher lost the ability to use his right hand at the age of thirty-six when his fourth and fifth fingers began to inexplicably curl under, and he couldn't control them at the keyboard. Fleisher, who died in 2020 at the age of ninety-two, was the most recognizable name in the piano world for decades, in part due to a career interrupted by the tragic loss of the use of his right hand, but perhaps more important, because he reinvented himself multiple times in order to follow his passion for making music. The title of his memoir says it all: *My Nine Lives, a Memoir of Many Careers in Music.*[28] It recounts the many reconfigurations of his career, from child prodigy to brilliant performer and award-winning pianist; from sinking to the depths of despair after developing a movement disorder in his right hand to becoming a virtuosic interpreter of left-hand repertoire, the most sought-after teacher in the country, a conductor who explored contemporary repertoire as well as the classics, artistic director of the Tanglewood Music Center, and finally returning to the concert stage as a two-handed pianist. Fleisher is seen in Figure 7.5 at the Music Teachers National Association Conference in Baltimore in 2017.

At the age of twenty-four, Fleisher became the first American to win the Queen Elisabeth Piano Competition in Brussels in 1952. In 1954, he collaborated with conductor George Szell on recording all the Brahms and Beethoven piano concertos, recordings that are still considered by many to be the finest recordings ever made of these works. In 1959, he accepted a position at the Peabody Conservatory, adding teaching to a heavy performing and recording schedule. But in the early 1960s, Fleisher began to notice that his fingers weren't responding quite right. "It seemed as if one or another finger was a little bit lazy."[29] His response was to practice even more, but eventually, the fourth and fifth fingers of his right hand began to cramp and curl towards his palm, and he couldn't control them in performance. He was scheduled to go on tour in 1965 to the Soviet Union with Szell and the Cleveland Orchestra but was forced to withdraw.

Focal dystonia was not understood well enough to be the first diagnosis on anyone's mind. The many doctors Fleisher consulted had no idea

Figure 7.5 Leon Fleisher, pianist
Credit: © 2017 Harry Butler, Nashville

what was causing the problem, although numerous diagnoses, including that it was psychological, were suggested. Fleisher's memoir and the many interviews he gave on the subject recount his despair, as well as the endless attempts to find a cure: hypnosis, traction, lidocaine injections, cortisone, rehabilitation therapy, acupuncture, carpal tunnel surgery, biofeedback, and more. He resisted playing the left-hand-only repertoire, much of which had been written for the pianist Paul Wittgenstein, who had lost his right arm during World War I. Fleisher didn't want to be known as a left-handed pianist any more than Evelyn Glennie wants to be known as a deaf musician.

But he eventually decided that making music was more important than playing the piano with two hands. He delved into the repertoire written for left hand alone, the Ravel Concerto for the Left Hand became his calling card, and several composers wrote new left-hand works for him. He increased his teaching load at the Peabody Institute. With composer and pianist Dina Kosten, he started the Theater Chamber Players in Washington,

DC, conducting and playing both contemporary and classical works. He became an increasingly well respected conductor, was a frequent guest conductor around the country, and in 1973, became associate conductor of the Baltimore Symphony Orchestra. But in the midst of all this activity, he continued to try to solve the mystery of his right hand.

Meanwhile in the research world, neurologist C. David Marsden discovered disturbances in agonist-antagonist muscle interaction in individuals with focal dystonia, the muscles often called flexors-extensors, indicating that focal dystonia was not psychological but had a physical cause.[30] In the early 1990s, Nancy Byl, a physical therapist who worked with musicians at the University of California in San Francisco, and Michael Merzenich, the neuroscientist who had shown in the 1980s that neuroplasticity could occur in the brains of adult owl monkeys, collaborated on a study that showed that excessive repetitive movement led to reorganization of the somatosensory cortex in monkeys' brains. Later, research at Johns Hopkins University demonstrated the same kind of abnormal somatosensory representations of the hands in human patients with dystonia.[31] Instead of separate finger areas in the somatosensory cortex, those areas overlap in musicians with hand dystonia.[32] Later, the representation of the lips in the somatosensory cortex was found to be altered in patients suffering from embouchure dystonia.[33] Focal dystonia was clearly not a psychological condition.

Focal dystonia and maladaptive neuroplasticity

Studies of musicians with focal hand or embouchure dystonia have found abnormalities, or maladaptive neuroplasticity, in one or more of three brain areas: reduced inhibition in motor areas, altered sensory perception, and impaired sensory-motor integration. Inhibition refers to the balance that is necessary in the nervous system between muscles that are firing and those that are not, allowing for smooth movement. When we speak of flexor and extensor muscles (for example, triceps and biceps), we're referring to activation of one muscle while the other is resting. This is called reciprocal inhibition— the process by which one set of a muscle pair is blocked so that the other can move freely. If playing something fast at the piano or on an oboe, while one finger moves, the others are inhibited, and the change between firing and inhibition must happen extremely quickly for all your fingers to be able to move independently. In dystonia, the muscle pairs try to work at the same time, one muscle does not get the signal to deactivate, so both are contracting. This results in a loss of control.[34] The same occurs in the muscles controlling the embouchure in embouchure dystonia.

A second abnormality is altered sensory perception. Musicians with focal hand or embouchure dystonia are unable to perceive two stimuli as spatially or temporally separate. Flutist Andrée Martin, diagnosed with focal dystonia, recounts her great surprise when a physical therapist asked her to close her eyes while the PT traced outlines of letters on the fingertips of both hands with a paper clip. Martin recalls that she had no difficulty whatsoever being able to tell what the letter was on her right fingers, but she was shocked to discover she couldn't identify the letters drawn on the fingers of her left hand.[35] When researchers have used brain imaging to look at the somatosensory cortex in musicians with dystonia, they have found that individual fingers overlap in musicians with focal dystonia and lip areas are altered in patients suffering from embouchure dystonia.[36] In Chapter 4, we saw that finger areas are enlarged in the somatosensory and motor cortices in skilled musicians as a result of practice (neuroplasticity). Increased sensory perception and motor control corresponds with becoming more skilled as a musician. But when neuroplasticity becomes maladaptive, those areas begin to overlap, and as with Martin, each finger loses its tactile ability.

Sensory-motor integration, the third possible area of abnormality, is the relationship between the sensory and motor system in our brains. As we learn an instrument, the brain learns to organize the various sensory inputs from fingers or embouchure and integrate them with the proper motor movement to create a certain kind or quality of sound. Performance improves as the sensory-motor integration becomes more efficient. But with focal dystonia, that integration can become impaired and sensory information no longer becomes connected to the appropriate motor area.[37] Eckart Altenmüller, director of the Institute of Physiology and Musicians' Medicine in Hannover, Germany, has researched and written extensively about dystonia in musicians. He has said that the newest theories point to focal dystonia as a larger network disorder. Not only can there be disturbances in the sensory and motor areas, but there can also be dysfunctions in connections between the basal ganglia, cerebellum, and premotor cortex (all have important roles in motor control). A person suffering from focal dystonia may have a disorder in any one or more of these areas, though not necessarily in all of them.[38]

Risk factors

There is no single cause for focal dystonia, but there are several factors that contribute to developing it. Males are affected four times more often than females, and focal dystonia usually appears in one's mid-thirties to forties, although it can begin as early as eighteen or as late as the mid-sixties. Musicians who develop dystonia have, on average, not begun music study until the age of ten, and musicians who have begun study when they are older than thirteen have five times higher risk of developing dystonia. Musicians who have begun study around the age of seven are at less risk because, when begun early, sensory-motor programs develop better and create a stable scaffold on which to build over a lifetime.[39]

Classical musicians are at greater risk than jazz musicians because in the process of learning a piece, they practice the same figurations repeatedly, trying to play perfectly what is in the score. Repetition, overuse, muscular tension, high motor demands, and the sometimes acute psychological stress related to fear of being judged when performing put classical musicians at greater risk for focal dystonia. Jazz musicians are less at risk because, although they may practice a significant amount, they are improvising so their practice routines are varied and with much less intense repetition and fewer constraints.

Still, only 1 to 2 percent of classical musicians develop focal dystonia despite repetitive practice, and there is increasing evidence that genetics may play a role. In about a third of cases, a family member also suffers from some form of dystonia, indicating a genetic predisposition, and researchers are increasingly looking into the genetics involved.[40]

Treatments

Dystonia symptoms rarely go away completely, and treatment is a challenge, but many musicians have been able to return to performing. The variety of treatments include oral medications—primarily trihexyphenidyl, injections with botulinum toxin, and rehabilitation or retraining techniques. Trihexyphenidyl blocks the chemical acetylcholine in the basal ganglia of the brain, a structure where automated movement patterns are stored, and thus allows movements to be corrected by conscious motor planning. In a study of sixty-nine patients, about a third showed significant improvement, but side effects were severe enough to make continuation of the treatment not possible.[41]

Botulinum toxin injected into the dystonic muscles has been successful in treating the symptoms of dystonia. It is incredibly toxic, but when used in tiny doses it can relax excessive muscle contraction. There is evidence that if patients injected with Botox practice while the muscles are in a relaxed state, that motion is stored in motor memory and, in some patients, can over-write the memories of the excessive muscle contraction.[42] A ten-year study of its use for focal dystonia patients found that it was safe and effective after a decade of treatment. Injections of Botox are given as needed, not sooner than three months and usually between three and six months. Professional musicians in the study usually timed treatments to obtain the best effect around scheduled performances.[43] Altenmüller has found it to be safe and effective for up to twenty-seven years of treatment (as of this writing) in some patients.[44]

Leon Fleisher was able to return to performing with both hands after treat-ment with botulinum toxin and Rolfing, a kind of bodywork that manipulates and reorganizes the fascia, or connective tissue, that supports bones and muscles. Fleisher was not the first to turn to alternative treatments for dys-tonia. 150 years before Fleisher, another well-known pianist was trying to find a cure for dystonia, although the condition did not yet have a name.

Robert Schumann's (1810–1856) hand issues are well known. Because of problems with the middle finger of his right hand, he was forced to give up his plans to become a concert pianist and instead turned to composing. Although that was heartbreaking for Schumann, it was fortunate for the world of music. Altenmüller has diagnosed Schumann's hand problems as focal dystonia. Looking specifically at entries in Schumann's diaries, let-ters, and physician reports during the years 1829–1833 when Schumann was developing hand problems, Altenmüller matched symptoms that were discussed to those of focal dystonia. Schumann tried various remedies, in-cluding adopting a different technique and using a stretching device, which unfortunately caused further problems. Altenmüller points to the well-known Toccata in C major, Op. 7, which can be played entirely without the middle finger, as evidence that Schumann was trying to adapt to playing without that finger.[45]

The third treatment for focal dystonia is rehabilitation or retraining techniques. Focal dystonia is not a degenerative disorder. It is a disorder of how nerve cells and connections function. And because maladaptive neuroplasticity is the root of that dysfunction, retraining to create new neural connections has been found to be effective. A pilot program of four patients in which fine motor control exercises were used away from the instrument resulted in improvement for all, with two of the patients being able to return to pre-dystonia performance levels.[46]

Musicians and coping with focal dystonia

Neurologist Frank Wilson, who celebrates the importance of the human hand in his best-selling book *The Hand*, reinterpreted Glenn Gould's professional and medical history and concluded that he also suffered from focal dystonia. Based on Gould's medical history, the print and film record, and Gould's unpublished 1977–1978 diary, Wilson argues that "in biomechanical terms Gould may have been almost completely unsuited for a career at the piano. Indeed, there is persuasive evidence that for virtually his entire career Gould struggled against and adroitly finessed critical limitations in upper body, forearm, and hand movements."[47] Wilson outlines comments made by Gould, the workarounds he made in order to be able to play, the physical crises that caused departures from the stage on two different occasions, and Wilson's own observations about what one can see visually in film clips, to make a persuasive argument for Gould's focal dystonia. While Schumann tried an external device and numerous other treatments, Gould appeared to have tried various workarounds at the keyboard to be able to continue to play.

Ironically, Wilson was the neurologist who diagnosed composer/pianist Jake Heggie's focal dystonia when Heggie began to develop the condition in the late 1980s. Heggie was in his mid-twenties at the time, which is early for developing focal dystonia. He had finished a degree in piano performance and composition at UCLA and had begun graduate studies in composition. He noticed that his right hand was cramping, and the right middle finger would pull under. He compensated by pulling up with the index finger, which caused the thumb to pull in. He was playing a great deal, making his living playing for soloists and choruses, and was beginning to make a lot of mistakes, so the cramping of his hand was of great concern.[48] A friend introduced Heggie to Wilson, who diagnosed early-stage focal dystonia and sent him to pianist Nina Skolnik at the University of California, Irvine. Skolnik had retrained many pianists with various kinds of injuries, including focal dystonia. She uses a combination of approaches including Taubman technique, Alexander technique, yoga, and other somatic (body) approaches. She works with the pianist on better alignment and balance, strategies to improve coordination, cooperative movement, and developing somatic awareness. Above all, she stresses that improvement can only occur when there is a change in the mindset or beliefs about the strategies or technical motivations that set the stage for dystonia to occur.[49]

For Heggie, that may have been the very curved finger position that he had learned, using his fingers precisely, rather than naturally. He worked with Skolnik for two and a half years, reworking his technique, going back to basics, learning to use rotation and the natural weight of his body rather than forcing, relaxing his finger position, and regaining facility at the piano. At that point, he moved to San Francisco and got a job with the San Francisco Opera writing press releases. He also began playing for singers. That's where the story begins that many people know. He showed one of his songs to mezzo-soprano Frederica von Stade, she loved it, and soon many other singers were asking Heggie to write songs. He was named composer-in-residence to the San Francisco Opera, asked to write an opera with the playwright Terrence McNally, and the result was *Dead Man Walking*, a powerful opera that, at the time of this writing, has had more than seventy productions. Many other operas, stage works, songs, song cycles, chamber works, choral works, and works for orchestra have followed.

Heggie considers himself fortunate to have "fallen into the hands of the right people right away" instead of having to spend years trying to get a diagnosis and treatment. Nonetheless, he says, "Your whole identity is wrapped up in the instrument you have studied your entire life. The desire to be a performer was still in there, so it was psychologically very challenging." But he thinks he wouldn't have had the same career had the focal dystonia not happened at that particular time.[50] Although he is best known as an opera and song composer, Heggie, shown in Figure 7.6, also frequently performs as a pianist with singers such as Frederica von Stade, Jamie Barton, Joyce DiDonato, Joshua Hopkins, and others.

Heggie was guided through a rehabilitation program by someone who had experience working with pianists with injuries. Other musicians have developed their own strategies for dealing with the disorder. Pianist James Litzelman, shown in Figure 7.7, didn't begin studying the piano formally until the age of sixteen, although he had taught himself how to play and arrived at his first piano lesson having written out several pages of George Gershwin's *Rhapsody in Blue* from listening to it multiple times. He had also

Figure 7.6 Jake Heggie, composer/pianist
Credit: © James Niebuhr

drawn the manuscript paper. He developed focal dystonia just fifteen years after beginning lessons, while he was finishing his doctoral degree at the Catholic University of America. Litzelman tried acupuncture and he tried Rolfing, neither of which helped, and then he discovered symmetrical inversion, which was his pathway to returning to performance.[51]

The human brain prefers symmetrical movement, as we saw in Chapter 5, with both hands playing in mirror image so the same muscle groups in each hand are contracting synchronously. These movements tend to be more

Figure 7.7 James Litzelman, pianist
Credit: John DeGrazia

stable, but that's not the way any instrument works. Usually, each hand performs totally different movements. We must suppress the brain's preference for symmetry through practice of asymmetrical movements. Symmetrical inversion is applicable only to keyboards, so discussion here will be brief.

Basically, it involves playing an excerpt of music in one hand with the symmetrical inversion of that excerpt in the other. The keyboard is symmetrical from two places: D and A-flat. If you begin with both hands on D and play an ascending D-major scale in the right hand and mirror it in the left, following exactly the same pattern of half and whole steps, you will have a descending G natural minor scale in the left, beginning and ending on the fifth degree of the scale. Different tonality, but same pattern of intervals. An excerpt of music for the right or left hand can be played in symmetrical inversion by the other hand. *[Demonstration videos by Litzelman of symmetrical inversion, including the example above, can be seen on the companion website at item 7.7; pianist Graham Fitch gives a master class on symmetrical inversion at item 7.8 ▶.]*

There are two reasons that this strange strategy works. The first has to do with the brain's preference for symmetry and mirror movement. The second is something called neural crosstalk (also called interlimb skill transfer), which basically means that when performing mirror or symmetrical movements with both hands, motor commands issued in one hemisphere spill over to the other via the corpus callosum.[52] One hand is teaching the other, or one hemisphere of the brain is teaching the other hemisphere. Interestingly, a 2014 study combined mirror hand movements with transcranial direct current stimulation (tDCS), which inhibited the part of the sensory motor cortex related to the dystonic movements while activating the healthy part of the motor cortex on the other side. This is basically copying the healthy motor program with the help of electrical stimulation to the hemisphere demonstrating dystonia.[53] Even without the electrical stimulation, Litzelman has been able to overwrite the dystonic movements with healthy movements by incorporating symmetrical inversion into his practice every day, and he has been able to return to an active performance schedule. Altenmüller cautions, however, that this is probably a very individual situation, as are many of these recovery programs that various musicians have tried. What works well for one musician with focal dystonia may not necessarily work for another.

Researchers are still learning a lot about focal dystonia. It has been only twenty-five years since it was discovered that maladaptive neuroplasticity plays a role. The role of genetics and network disorder is much more recent. But since there are so many contributing factors, each individual's experience and road to recovery or acceptance is different. However, a common experience recounted by everyone who has received a diagnosis of focal dystonia is a profound sense of loss, often accompanied by deep depression. When you have spent the majority of your life, and all your adult life, consumed with making music, facing the thought of not doing so feels as though you have lost your identity, you have lost who you are. Many recount a period of despair before they begin looking for answers. They try numerous medications and therapies and consult practitioners of alternative therapies until they find what works for them. As trombonist David Vining has said, "don't wait for someone to present you with a magic bullet. Rarely is there a therapy template for recovery. Question traditional pedagogy and look at your recovery process as a unique cocktail of therapies. It's up to you to pull them together in various ways until you find a way that works for you."[54]

And that's what Vining did. Most doctors and researchers say that embouchure dystonia cannot be effectively treated, but Vining, shown in Figure 7.8, has shown that this is not the case. Shortly after beginning a new position at the University of Cincinnati College-Conservatory of Music, he began to develop dystonia symptoms. In an essay titled "Why Don't You Just Play the Saxophone?" (which is what one doctor said to him), Vining talks about the many doctors he consulted, the importance of being your own advocate, and the importance of cultivating a sense of global awareness so that you are looking at your entire body, not just the dystonia. Understanding that his problem was caused by faulty neuroplasticity, he decided to use neuroplasticity to create new pathways in his brain, even though he was told that couldn't be done. He describes how he used the concept of neuroplasticity to guide his recovery, establishing new pathways for control of his embouchure in his brain by building from the bottom up—relearning

Figure 7.8 David Vining, trombonist
Credit: Photo used by permission of David Vining

how to move air through his instrument instead of concentrating on his embouchure. His essay is contained in *Notes of Hope*, a compilation of several stories by musicians who have recovered from various injuries.[55] *[A link to Vining's essay can be found on the companion website at item 7.9 ▶.]*

Alex Klein is another musician who devised his own recovery. Klein, seen in Figure 7.9, began playing the oboe at the age of nine in his native Brazil. He progressed quickly, and after earning a bachelor's degree and artist diploma from Oberlin, he won several competitions and joined the Chicago

Figure 7.9 Alex Klein, oboist
Credit: Todd Rosemberg

Symphony Orchestra (CSO) in 1995 at the age of thirty. He won a Grammy in 2002 for his recording of the Richard Strauss Oboe Concerto with conductor Daniel Barenboim and the CSO. He was in a job he loved, had a lot of performing opportunities, and was seemingly at the height of his career. But he noticed in the late 1990s that the third and fourth fingers on his left hand weren't working quite right. By 2001, it had become bad enough that he consulted several doctors, including Alice Brandfonbrener, the founder and director of the Medical Program for Performing Artists in Chicago and one of the founding members of the field of arts medicine. She diagnosed his problem as focal dystonia and told him he would probably have to leave the orchestra within three years.[56]

And in fact, that's what happened. But not before he had tried chiropractic, acupuncture, physical therapy, the Alexander technique, and hand massage. He had a new, lighter oboe made. He tried Botox, levodopa, finger weights. The changes he made in his playing to try to accommodate the dystonia led to painful tendinitis. By 2004, he asked to resign from the CSO because he wasn't able to play at the level he thought the orchestra deserved. He moved back to Brazil suffering from extreme depression. But eventually, as Leon Fleisher had done, he found other means of musical expression. He took up conducting, founded a major music festival, and started a Brazilian orchestra for at-risk schoolchildren inspired by the El Sistema program in Venezuela.[57] But he didn't give up his dream of playing with a major orchestra. He learned how long he could practice before encountering problems.

Since focal dystonia is task-specific, it affected only his oboe playing. He could still play oboe d'amore, English horn, or baroque oboe. So he tried what is known as a "sensory trick." He glued a small Brazilian coin over the G-key of his oboe, providing an extension for his fourth finger. That meant his hand was playing in a slightly different position, which his brain didn't recognize so it didn't cause his finger to cramp.[58] He changed his technique somewhat to rely more on his right hand, and when he plays with his left, instead of extending his fingers fully to reach the keys, he rotates his arm.[59] The oboe Klein is holding in Figure 7.9 was specially made for him by the F. Lorée company. It has an extended G-key, with other tone holes and keys adapted so they can reach out toward his hand.[60]

Klein found a way to approach playing that worked for him. He is now principal oboe with the Calgary Philharmonic Orchestra in Canada, teaches at DePaul University in Chicago and at the Aspen Festival, and in 2020 released a highly acclaimed CD on Cedille Records of Twentieth Century Oboe Sonatas with pianist Phillip Bush.

Not every musician who develops focal dystonia is able to return to the performance level at which he previously played. But as David Vining has said, creativity is key. For the musicians who are willing to explore alternative options, to throw out everything they thought they knew about playing their instrument, who are willing to take chances, a return to playing is often possible. Some things may be better. Vining reports a new ease in playing, as does Jim Litzelman. But the focal dystonia never totally goes away. Litzelman says that performance anxiety can cause some symptoms to return; Klein has said that playing too long brings on symptoms, so he has to limit his playing time. Heggie says that if he looks at music that he used to play pre-dystonia, he begins to feel his hand changing. Still, as these several musicians show, a return to performance is possible for some musicians.

Nina Skolnik says that very often, a pianist with dystonia "confesses to 'perfectionism,' a 'mind over body' ethic with the underlying affirmation 'I have enough *will* to conquer this.'" But recovering from focal dystonia isn't done with *will*. "Pushing through it" is not an option. Changing one's mindset is crucial to recovery.[61] Eckart Altenmüller has commented that "if you are extremely driven to recover and if you are extremely focused on just getting rid of the symptoms of dystonia, you will not be successful. You must change something about your life, the overuse, too much practice. People who are successful in retraining changed this kind of abuse of themselves."[62]

In an article Klein wrote for the journal *The Double Reed*, he gives advice to those encountering focal dystonia.[63] *[A link to the full article can be found on the companion website at item 7.10 ▶.]*

Take the curve on the road and accept it. The plans, dreams and expectations you had for the previous direction on the road are now lost, forever. Accept it. Open your hearts to the new challenges, which promise to be every bit as interesting as the older ones, if not more. It might just happen that by accepting the unacceptable you will place your mind at ease enough to find solutions you would never have considered had you maintained your mind set on your past. And it might just be that some of that past will return to you on a silver platter, as a new beginning.

Key Concepts

- Passion drives neuroplasticity. The more motivated one is to learn, the more readily neuroplasticity occurs.
- Neuroplasticity makes it possible, in some cases, for a brain-damaged musician to return to performing by creating new neural pathways in the brain.
- Blind or hearing-impaired individuals can become superb musicians due to cross-modal neuroplasticity, in which the visual cortex responds to signals from touch or hearing, and the auditory cortex responds to touch signals.
- Neuroplasticity sometimes goes awry, as in focal dystonia, a large group of movement disorders that are characterized by uncontrolled movements and involuntary muscle contractions. It is a network disorder that can involve abnormalities in any one of several brain areas.
- Although musicians suffering from focal dystonia are often told by doctors that they will never play again, many have used a mix of creativity, perseverance, and an openness to all possibilities to devise their own therapies or mix of therapies—ultimately regaining playing ability and a successful return to performance.

8

Imagery—Music in the Mind's Eye, Ear, Body

Keep searching for that sound you hear in your head until it becomes a reality.

—jazz pianist Bill Evans

Fei-Ping Hsu's story

It takes years of practice every day to become a world-class musician. So how does one explain a young Chinese piano prodigy who at the age of fourteen was prevented from practicing for ten years, yet went on to have an international career? Fei-Ping Hsu was one of the first student musicians to be allowed out of China to study in the West following the end of the Cultural Revolution. He played exquisitely, which was baffling to anyone who discovered that he had not been able to practice during the entirety of that period—from 1966 to 1976. I met Fei-Ping at the Eastman School of Music and heard him perform several times during the year he was in residence there. He had a brilliant technique, but also a lush sound filled with emotion, nuance, and sensitivity. I was increasingly astonished as I learned more about his recent past. It was difficult to believe that he could play at all, let alone perform so beautifully. At that time there seemed to be no accounting for this feat. But discoveries about the brain in the 1990s do provide an explanation for his remarkable story, which is worth telling in some detail. (The account in this chapter is based on my conversations with Hsu during 1979 and 1980, with his brother Fei-hsing Hsu and niece Hsing-ay Hsu in 2015 and 2021, and from *In Memory of Fei-Ping Hsu: A World Classic Pianist*, edited by Yiwan Peng;[1] text in Mandarin Chinese, personal translations by Anne Pusey.)

Hsu was born in 1952 on the beautiful island of Gulangyu, a tourist destination in southern China known for its beaches. It is often called "Piano Island" or "Music Island" because the Christian missionaries who were there

The Musical Brain. Lois Svard, Oxford University Press. © Oxford University Press 2023.
DOI: 10.1093/oso/9780197584170.003.0008

in the early 1900s not only introduced Western music to Gulangyu but left behind a significant number of pianos on the island. Hsu's family had converted to Christianity and the hymns his grandmother and mother played in church were his introduction to Western music.

His older brother Fei-hsing Hsu, also a pianist, recalls that at the age of six, Fei-Ping filled in for his mother at church, playing the hymns by ear. The family recognized his talent and arranged for him to study with a well-known singer who lived on the island. But when the head of the piano department at the Shanghai Conservatory, Ji-sen Fan, visited the island and heard Fei-Ping, he was so impressed he took him as a student. Fei-Ping was then eight years old, the youngest student ever to study at the conservatory. Fei-Ping had exceptional facility and musical depth, and by the time he was eleven, he was playing all twenty-four of the Chopin études, a difficult feat even for far older pianists. At twelve, he performed for Queen Elisabeth of Belgium when she visited China. She was so captivated with his talent that she invited him to come to Europe to study and offered to help him get governmental permission.

Russian visitors to Shanghai were also impressed because he played in a Russian style, with a big, emotional sound, not the cooler sound usually associated with Chinese musicians. This was perhaps not surprising since his teacher at the conservatory had studied with Russian pianists. These international connections might have moved Fei-Ping's career in an exciting direction, but this was 1966 and one beautiful day when the fourteen-year-old arrived at the Shanghai Conservatory as he did every day to practice, study, and take classes, he and his fellow students found the pianos missing from the practice rooms. The instruments were in the dining hall dumped in a heap. The teachers were gone, the valuable scores and recordings destroyed. China's Cultural Revolution had begun.

Communist Party Chairman Mao Zedong (Mao Tse-tung) set the Cultural Revolution in motion to purge capitalist elements from Chinese society and get rid of the "Four Olds": old customs, old culture, old habits, and old ideas. That meant attacking and destroying cultural institutions and treasures, books, music, and art. Western music and art were considered part of the old culture and therefore an "enemy of the people." Many teachers and scientists were persecuted and killed; students were expected to denounce their former teachers, something Fei-Ping refused to do. He continued to revere his teacher, Ji-sen Fan, and regularly visited him, even though it was dangerous to do so. Practicing was forbidden, and even listening to Western music was banned.

All young people of student age were sent to the country to work in the rice fields. The exceptions were some musicians, such as Fei-Ping's older brother, who were required to perform revolutionary songs in factories and at political meetings. The "Down to the Countryside Movement" was Mao Zedong's policy to send what he considered privileged urban students to learn from the rural peasants—"revolutionary education." Kneeling over and picking kernels of rice in the cold water was very hard on Fei-Ping's hands and fingers, and the young student workers were given little to eat other than rice. After a time, Fei-Ping was sent to a factory where he carried heavy metal machinery parts on his back. He thought about trying to escape—but where could he go? All of China was suffering from the same conditions. The years were going by and all he could think about was music. He saw his musical future disappearing.

While Fei-Ping didn't talk a great deal about the Cultural Revolution during his year at Eastman, he did say that during all those bleak years, he would listen in secret at night to Russian recordings of Horowitz and Rachmaninoff. This was illegal and he would have been severely punished had he been caught, but he was desperate and somehow managed to escape detection. He would imagine himself playing the works he was hearing, feeling in his shoulders, arms, and hands what it would be like to play the music.

Around 1969, Fei-Ping had a bit of a reprieve when someone remembered his prodigious talent, and he was called to Beijing to be the back-up pianist for the Yellow River Piano Concerto with the Central Philharmonic Orchestra Association in Beijing. The concerto was based on the Yellow River Cantata, written in 1939 and based on traditional folk melodies and the image of the Yellow River. One of the longest in China, the Yellow River was revered as a symbol of the courage of the Chinese people. The concerto combines elements of Chinese folk music with traditional European classical music (somewhat ironic, since Western music had been banned in China). It has a virtuosic solo part and has become the most popular piano concerto in China.

The first pianist of the orchestra was a member of the Communist Party. Fei-Ping did not belong to the party so, as second pianist, he had to be on call to play at the last moment whenever the first pianist was not available. He was still not allowed to practice. The Yellow River Piano Concerto is fiendishly difficult for the pianist, and pianists who perform it today, such as Lang Lang or Yuja Wang, would never consider playing it with no practice, but Fei-Ping had no choice. Playing difficult repertoire in concert without practice over a period of years takes a tremendous physical toll on the body. After two years of being on call to play the concerto at a moment's notice, Fei-Ping had developed what was probably a stress injury, and he asked for time off to go to Shanghai for hand therapy. The Communist Party allowed him six months. At the end of that period, he was called back to Beijing on one day's notice to again play the concerto—having had hand therapy, but no practice during the interim.

He played one performance and was then out of a job, and he struggled until the end of the Cultural Revolution in 1976. Fei-Ping was fourteen when the Cultural Revolution began, twenty-four when it ended. These are formative years for a young artist, when technique is developed and the foundation laid for the rest of one's performing life. Even after the end of the Cultural Revolution, there were no teachers, and the only pianos were in the dorms at the Shanghai Conservatory. They were in poor condition, but Fei-Ping was able to do some practicing there. However, his primary method of engaging with music during those years, as it had been during the entire Cultural Revolution, was in his mind.

Luckily for Fei-Ping, well-known Chinese bass-baritone Yi-Kwei Sze visited Shanghai in 1978. Sze had left China in 1947 to pursue a career in the United States, had a highly successful concert and operatic career, and was on the voice faculty at the Eastman School of Music. Sze had gone back to China to visit friends and family as soon as international relations began to thaw. While Sze was teaching at the reopened Conservatory, Fei-Ping came to his attention. Sze arranged for Fei-Ping and two other students to study at Eastman.

I wasn't alone in wondering how Fei-Ping could perform with such sensitivity, grace, and power. Everyone who heard him was astonished at this slight man who performed with such emotional intensity. But discoveries in neuroscience in the 1990s began to shed light on Fei-Ping's remarkable success. It became clear, in light of new research on *motor imagery*, that making music in the mind, with the "mind's muscles," as Fei-Ping had been doing when he listened to recordings of Horowitz and Rachmaninoff, was indeed practicing.

What is imagery?

If asked to imagine a favorite vacation spot, most people will have no difficulty doing so. You may visualize lying on a beach or hiking your favorite forest trail. You can *see* those images in your mind, and they are a welcome escape when you're feeling burned out. You may also mentally *hear* the waves crashing on the shore or the wind in the trees, but you probably won't *feel* the sand between your toes or *feel* what it's like to walk on a trail covered with pine needles. If you do, you're in the minority. As you read this, you may try feeling sand under your feet and say "Sure, I can do this," but it doesn't tend to come as fast or as easily as auditory or visual imagery. (There is a very rare condition called *aphantasia*, either acquired or congenital, in which an individual is unable to form mental images. Some people who suffer from aphantasia are also unable to imagine sounds.)

Imagery is the subjective experience of seeing, hearing, or feeling in one's mind without any actual sensory stimulation. Although "image" is synonymous with "picture," imagery does not refer only to the visual. Imagery accurately refers to any of the senses and includes physical sensations and emotions as well. Musicians tend to use visual and auditory imagery frequently. Even lacking an eidetic (photographic) memory, many musicians see the score in some form in the mind, perhaps where the page breaks are or where on the page one section ends and another begins. A musician might visualize the keyboard layout, keys on a woodwind instrument, or a string instrument's fingerboard. One can visualize the performance space, walking onstage, and even seeing an imaginary audience. (One of the tricks that has been used for years to alleviate performance anxiety is to imagine the members of the audience without clothes.)

Musicians are also accustomed to using auditory imagery, hearing music in the "mind's ear." Sometimes a piece of music being practiced will lodge itself in the mind and continue playing endless repetitions throughout the day—that's called an *earworm*. Although musicians talk about and use auditory and visual imagery, they are far less likely to use motor imagery—imagining or feeling movements in the mind. Just as people don't tend to imagine how their bodies feel in the salty water at the ocean, musicians don't spend much time imagining how their bodies feel when playing a particular piece of music. Singers are more likely to imagine how their bodies feel, because bodily sensations are so closely tied to technique.

Feeling in the mind's body is motor imagery, imagining motor movements without physically making them. There are five commonly acknowledged senses: vision, hearing, touch, taste, and smell. Proprioception, as we saw in Chapter 4, is considered the sixth sense. But there is another sense, and that is *kinesthesia*, the sense that delivers information to the brain about effort, movement, position, and weight delivery (we talked about developing one's kinesthetic sense in Chapter 6). It is basically the ability to sense the motion of a joint or limb. Most people don't pay attention to their movements—limbs and body are expected to move, unless one is recovering from a broken bone or other accident. But we do have the ability to pay attention to our motor movements and that is kinesthesia.

Kinesthesia is sometimes used interchangeably with proprioception, but they are not the same, and they are referred to by some as the sixth and seventh senses. *Proprioception* is about one's sense of self, an inherent sense of one's position and motion in space. If you hold your arm above your head or behind your back, you can't see it, but you know exactly where it is. That's proprioception. Kinesthesia, on the other hand, is knowing how it *feels* to move your arm above your head or behind your back. Kinesthesia is the sense we use when we practice *motor imagery*, imagining a physical movement in our minds without executing it. The term motor imagery is sometimes used interchangeably with kinesthetic imagery. When we imagine, or try to feel in our body, arms, and fingers what it is like to play a piece of music—just as Fei-Ping Hsu did—that's motor imagery.

Composers, performers, and imagery

Many composers have spoken about the use of imagery, or before that word was in use, of composing or hearing "in the mind." In the late eighteenth century, Mozart wrote, "the whole, though it be long, stands almost complete and finished in my mind, so that *I can survey it, like a fine picture or a beautiful statue, at a glance*. Nor do I hear in my imagination the parts *successively*, but I hear them, as it were, *all at once*. . . . For this reason, the committing to paper is done quickly enough, for *everything* is, as I said, already finished, and it rarely differs on paper from what it was in my *imagination*."[2]

The great works of Beethoven's last ten years, including the final piano sonatas and string quartets, the Ninth Symphony, the Missa solemnis, and the Diabelli Variations were written when he was profoundly deaf, hearing these works only in his mind. Beethoven told his first biographer, Johann Schlosser, "The working out in breadth, length, height, and depth begins in my head, and since I am conscious of what I want, the basic idea never leaves me."[3]

Robert Schumann (1810–1856), pianist, music critic, and one of the greatest composers of the nineteenth century, wrote often about imagery in his *Advice to Young Musicians* in 1850. The translations below come from cellist Steven Isserlis's wonderful volume in which Isserlis adds his own commentary to each of Schumann's gems:[4]

It is not only your fingers that should know your pieces; you should also be able to hum them away from your instrument. Sharpen your imagination to the extent that you remember not only the melody of a composition, but also the harmony that belongs to it.

It has been said that a perfect musician must be able to visualize even a complicated piece of orchestral music on first hearing as if it were lying before him in full score. This is indeed the greatest achievement that can be imagined.

American composer Henry Cowell wrote: "the most perfect instrument in the world is the composer's mind. Every conceivable tone-quality and beauty of nuance, every harmony and disharmony, or any number of simultaneous melodies can be heard at will by the trained composer." Hearing in his mind didn't come easily for Cowell and he spoke about having to practice hearing sounds in his mind. But he relates that he practiced until he could hear music in his mind with ease.[5]

Composer Jake Heggie, who has been called "arguably the world's most popular 21st century opera and art song composer" by the *Wall Street Journal*,[6] uses both visual and auditory imagery when he composes, but visual is primary. Heggie says, "when I know what I am writing about, I visualize the person I am writing for, the situation they will be in, whether it is a character or it is a singer I know who will play that character, and then I start to hear it, and it starts to come to life for me."[7] For a composer who has had over seventy national and international productions of his opera *Dead Man Walking*, and more than twelve productions of the opera *Moby Dick*, perhaps it's not surprising that the visual would play such a prominent role in his compositional process.

Performers have often used imagery as a tool for learning. Walter Gieseking (1895–1956), well-known pianist and teacher, taught visual imagery as part of his teaching technique to aid memorization, to better understand the structure of the piece, but chiefly to "train the ear."[8] Gieseking was well known for memorizing while traveling by train or plane. Rosina

Lhévinne, inspirational pianist and teaching legend at the Juilliard School, would tell her students, "You imagine the sound you wish to produce, and then you produce it."[9] Pianist Glenn Gould calculated that "at the age of twenty-seven . . . he had played the fifth Bach Partita roughly five hundred times, mostly while driving or walking around town."[10]

It's not only pianists and composers who use imagery; so do singers. Soprano Lynn Eustis, who has sung over thirty operatic roles and soloed in numerous oratorios and other choral works, says she *audiates* an entire recital, in real time, before singing it, and encourages her students at Boston University to do the same. For Eustis, this means not only hearing the music in her mind, but also hearing the text in her mind (including foreign languages and the translation), imagining the tone she wants to produce and the placement of that tone, and hearing the piano or orchestra with which she is performing. If she's singing an opera, she hears every other role in her mind as well.[11]

Athletes and imagery

Athletes have been using imagery or visualization at least since the 1950s. Pancho Gonzalez, legendary American tennis player in the 1960s, was said to have visualized every match he played. By the time he went on court, he had already won the match in his mind. Billie Jean King, another tennis great and winner of thirty-nine Grand Slam titles in the 1960s and 1970s, always used visualization before a match, thinking of everything that could go wrong and imagining how she would handle it, whether the situation was rain, wind, or how the ball might be coming at her. She even pictured where she wanted the ball to go as she hit it on the court.[12]

In the mid-1980s, researchers in Canada decided to look seriously at the importance of mental training, and who better to study than Olympic athletes. They interviewed or sent questionnaires to 235 athletes who had participated in the 1984 Sarajevo Winter Olympic Games and Los Angeles Summer Games. The athletes were questioned about their physical, technical, and mental readiness, their background in mental training, the role of coaches and sports psychology consultants, and more. The sample included thirteen Olympic gold medal winners, three silver, one bronze, and three world champions. The subjects were forthright about the kinds of mental training they engaged in. A diver commented, "I did my dives in my head all the time." A pistol shooter: "I actually shoot in imagery. . . . I can feel the initial pressure of the trigger." A figure skater: "My imagery is more just feel. When I'm actually doing it I get the same feeling inside. You have to experience it, and once you do, then you know what you are going after."[13]

The athletes spoke about perfecting imagery through daily practice: "It took me a long time to control my images and perfect my imagery, maybe a year, doing it every day." Many athletes expressed the belief that they could have reached the top much earlier if they had strengthened their mental skills—their technique was the same the second time at the Olympics, but the mental training took them over the top. Ninety-nine percent of the 235 athletes in the study reported using mental imagery. They used it at least once a day, four days a week, for at least twelve minutes and up to two to three hours right before the actual Olympic event. These athletes said it was important to "feel" in imagery as they did when they performed, and the ease with which they could control the imagery correlated with successful Olympic performances. The study had looked at three major factors impacting performance: mental, physical, and technical. Mental readiness was the only one that had a statistically significant link with their final Olympic rating.

Olympic athletes today still use imagery extensively, but they think of it as multisensory—using all their senses and prefer "imagery" to "visualization." Emily Cook, the American aerialist who competed in the 2006, 2010, and 2014 Olympics, says, "You have to smell it. You have to hear it. You have to feel it, everything."[14] Cook began using imagery in 2002 after she suffered broken bones in both feet when she crashed during a training jump in Lake Placid, New York. Her injuries kept her from skiing for more than two years, but during that time, she and her coach used imagery. They created scripts that Cook would record and play back, scripts that engaged all her senses—hearing and seeing the crowd, feeling the wind on her neck, engaging her core, going through and feeling in her body every step of the jump. While listening to the imagery scripts, she would feel her muscles firing in response.[15] She was engaging in visual, auditory, and motor imagery. She went on to be a six-time US National Champion in Aerial Skiing, as well as competing in three Olympics. What's important here is that Cook wasn't just mentally "seeing" every facet of the jump; her body was "feeling" it, just as the athletes from the 1984 Olympics spoke about "feeling" in their bodies.

Terminology matters

With both athletes and musicians, the terms "imagery" and "imagination" are often replaced by "mental practice," which is far less specific. The terms "mental practice," "mental training," "mental rehearsal," "visualization," and "imagery" are often used interchangeably, with no clear indication of what is implied. Scientists sometimes use mental practice to refer to auditory imagery, sometimes to motor imagery, and sometimes to both. Musicians may use mental practice to refer to the techniques used to manage performance-related stress and anxiety, to developing skills to maintain concentration and focus, and to handling the mental challenges that may be associated with injuries. Mental practice is also used to refer to silently analyzing a score, listening to recordings of pieces you are working on, visualizing walking onstage and performing, relaxation techniques, and hearing the music in your mind.

Imagery, whether auditory, visual, or motor, is a brain process. Areas of the brain have been identified that are involved in processing each of these kinds of imagery. Mental practice doesn't refer to a brain process; it refers to a cognitive process that may or may not include imagery. Mental practice as used for relaxation techniques, setting goals, techniques to focus attention, and similar strategies is extremely important, but unless it is being used to refer to imagery, *it does not change the areas of the brain that have to do with playing an actual piece of music on your instrument*. This chapter deals specifically with imagery, the brain areas that process imagery, and the impact of imagery on learning and performance.

Visual imagery—"seeing with the mind's eye"

Many people advocate visualizing a successful performance because, they say, the brain doesn't distinguish between what is real and what is imagined. The concept is that the more frequently successful performances are imagined, the more the brain will think they actually happened. But unless you are hallucinating, the brain does, in fact, know the difference between real and imagined visual experience. Visual *perception* is the brain's ability to make sense of the information coming through the eye. Visual *imagery* is based on memory of things one has seen. During imagery, one either experiences a version of the original perception from memory or creates in the imagination new versions of visual scenes based on familiar circumstances. A concert hall that doesn't exist can be visualized or imagined based on elements known to be in a hall: the stage, the seating areas and balconies, backstage, exit doors, etc.

Although the brain knows the difference, imagery and perception do share processing areas in the brain. They share a large overlap in neural processing in the visual, parietal, and frontal cortices.[16] Some researchers suggest that visual imagery functions as if it were a weak form of visual perception.[17] There is also an overlap between visual imagery and visual working memory.[18] Working memory is used to recall and recombine images from memory when engaging in visual imagery. Set design, mechanical engineering, photography, and graphic design all rely on good visual imagery and visual working memory.

Visual imagery has many advantages. Many teachers advocate imagining the hall in which a performance will take place as well as the entire routine leading up to performance. Because visual perception and imagery share so many processing areas, this will indeed make you feel as though you have done this before, as many Olympic athletes have stated. The more vivid the imagery has been, the more real it will seem. Although your brain knows the difference, you develop a comfort level with the performance, which has a positive effect.

But visual imagery can also have a negative impact if one worries about and visualizes negative outcomes. Studies show that visual imagery plays a pivotal role in clinical disorders such as anxiety.[19] Jacqueline Hernandez, an American Olympian in snowboard cross, related trying to conquer images of crashing that would pop into her mind. Nicole Detling, a sports psychologist who worked with the US Freestyle Aerials Ski Team at the Sochi 2014 Winter Olympics, says that "In images, it's absolutely crucial that you don't fail. . . . So one of the things I'll do is if they [the skiers] fail in an image, we stop, rewind and we replay again and again and again."[20] Emily Cook, the aerialist, relates using imagery to break a cycle of negativity. Whenever fear surfaced, she would imagine herself pricking a big red balloon with a pin, and that would snap her out of the negativity.

In addition to using visual imagery to imagine performances, one can, as Gieseking did, also use visual imagery to study scores. What one sees is transferred into working memory. All or parts of the score can be visualized. The more practiced one becomes at visualization, the more detail one is able to visualize in the score. Musicians who learn by visualizing the score will also be using auditory imagery to hear the music in their minds.

What visual imagery or visualization by itself cannot do is make any difference in the motor movements necessary to perform the piece. Visual imagery accesses visual processing areas. It has nothing to do with the movement of muscles or motor processing areas of the brain—unless one is also imagining the movement, and that is motor imagery.

Auditory imagery—"hearing in the mind's ear"

Most musicians regularly use auditory imagery. When not at an instrument, a musician will often hear the music in his mind that he is currently studying. Silently reading musical scores and hearing them in the mind is *notational audiation*. When sight-reading, one imagines the sound of the music at the same time as it is being played. Edwin Gordon (1927–2015), American music educator, researcher, author, and teacher, spent his career advocating for hearing "in the mind's ear." He coined the term *audiation* in 1975 to refer to internal comprehension of music, to hearing or feeling sound when it is not physically present. As you read these words, you are no doubt hearing them in your mind and giving them meaning. Gordon said the same is true with music. Audiation is not just about hearing music in the mind, but about giving it meaning and understanding.[21] We think of the ideal sound we want in performance and try to achieve that as we are playing. Barry Green wrote in *The Inner Game of Music*, "Effectively, you are playing a duet between the music in your head and the music you are performing."[22]

Some researchers use the term *audiation*, some refer to auditory imagery. Most research in auditory imagery focuses on the details of sound, such as pitch, timing, and timbre, not on the totality of a musical piece as musicians experience it. But researchers have found that, just as activity can occur in the visual cortex in the absence of seeing an actual image, neural activity can happen in the auditory cortex in the absence of sound. Pitch, timing, and timbre are processed similarly in the brain whether we are hearing the music or imagining it.[23] Emotional responses to music are processed similarly in the brain, whether the music is heard live or imagined.[24]

Many brain areas are involved in auditory perception and imagery including the auditory cortex, parietal and pre-frontal cortices, pre-motor and supplementary motor areas, and the cerebellum.[25] As with visual imagery, working memory is involved in auditory imagery, and there are interactions between the pre-frontal cortex (working memory) and the auditory cortex.[26] With auditory imagery, we recall music that we have heard, we hear music in our mind that we see on a page of music notation (notational audiation), or we create new music in our minds based on either intuitive or learned knowledge of melody, rhythm, and harmony.

Why are motor areas active when you are hearing music in your mind? Two theories have been proposed: (1) you may be thinking in terms of playing your instrument, but if that's the case, then you are also using motor imagery; and (2) you may be "subvocalizing" or humming internally. In either event, there is less motor area involvement than when using motor imagery.

In the world of brain and music research, singers haven't been studied as often as pianists or other instrumentalists because singing involves text and text includes brain processing areas for language in addition to music. But one interesting study involving singers was done in Germany in 2007. Sixteen classically trained singers, eight professional opera singers and eight students, had their brains scanned via fMRI while they sang and imagined singing excerpts from the well-known aria by Tommaso Giordani, "Caro mio ben." When the researchers compared brain activity from the actual and imagined singing, they found a lot of overlap in areas related to motor processing and auditory processing. During imagined singing they found increased involvement of fronto-parietal areas, which have to do with working memory. Interestingly, they also found more involvement of emotion processing areas during imagined singing than during actual singing, which means the singers were able to use emotion more freely when they imagined singing the aria.[27]

In 2017, soprano Renée Fleming partnered with the Kennedy Center and the National Institutes of Health to explore what was going on in her brain while she was singing, speaking, and imagining singing. She sang several times an excerpt from the folk song, "The Water Is Wide," while the researchers looked at her brain activity. Her brain showed more activity during imagining singing than either during singing or speaking, and Fleming commented that actual singing was easier. Imagining singing took more effort—more focus and intention.[28] *[A link to a clip of Fleming discussing this study can be found on the companion website at item 8.1 ⊕.]*

Auditory imagery or audiation is an acquired skill. While Mozart seemed to have had a natural ability for hearing his compositions in his mind before writing them down, he also lived at a time, as did Beethoven, when music was learned by ear rather than by reading musical notation. Other composers, including Henry Cowell, had to work at hearing music in their minds. Cowell was eventually able to do so. But being able to imagine music in the mind is not just a handy tool for composers or performers, it is necessary to be a good musician, as countless performers, composers, and music educators have indicated. Renowned pianist and teacher Leon Fleisher often commented, "You have to hear it before you can play it." Without an internal image of how we want a piece to sound, we won't be able to achieve that sound. Or as American jazz pianist and composer Bill Evans said in the quote at the top of the chapter: "Keep searching for that sound you hear in your head until it becomes a reality."

Learning to audiate or hear and understand music in our minds begins with singing and listening to music as a child. Gordon's *Learning Sequences in Music* is an excellent resource to help in understanding audiation; an introduction to learning sequences in pitch, rhythm, and patterns; and a guide to age-appropriate musical activities that lead a child toward musical understanding.[29]

Motor imagery—"feeling in the mind's body"

Motor imagery uses the sense of kinesthesia, the sense of position and movement. During motor imagery, we imagine, without moving, all the physical movements and sensations necessary to perform a piece. In fact, moving fingers, lips, or arms while engaging in motor imagery defeats the purpose because that activates different muscles, not the ones used in performance. All movements involving the motions used in a piece must be in the imagination.

We previously looked at a study in which right-handed adults who did not play a musical instrument practiced a five-finger exercise for two hours a day for five days (see Chapter 4). A baseline measurement of the right-hand finger area in the brain was taken using transcranial magnetic stimulation (TMS), then measured again after each two-hour practice session. In a confirmation of neuroplasticity, the cortical map corresponding to the fingers of the right hand increased in size each day.

Alvaro Pascual-Leone, the author of that study, began to wonder what would happen if the subjects practiced using only motor imagery. He decided to repeat the experiment, but this time the participants would replace physical practice with motor imagery, imagining the movements necessary to play the five-finger exercise. A different group of participants was assigned to one of three groups: a physical practice group, a mental practice group, or a control group.

The physical practice group practiced the five-finger exercise two hours daily for five days, just as in the first study—C D E F G F E D C—at a metronome marking of 60 corresponding to the C and G (thumb and little finger). The mental practice group imagined the movements necessary to play the pattern for two hours daily—and were monitored to see that they did not actually move their fingers. The control group didn't practice. As in the first study, the area of the motor cortex corresponding to the finger area of the right hand was mapped using TMS (transcranial magnetic stimulation) each day thirty minutes after practice.[30]

After five days, there was of course no improvement in the control group that hadn't practiced. The group practicing motor imagery, however, showed significant improvement in *performance* of the five-finger pattern, although not quite as much as the physical practice group. But the most interesting finding was that the motor imagery group *showed the same plastic changes in the motor system in the brain* as was shown by the subjects who had done physical practice, even though their performances were not quite at the same standard. And after a single two-hour physical practice by the mental practice group, their performances surpassed those of the physical practice group.

The same brain areas are involved in motor imagery as in actual physical practice, with the single exception of the primary motor cortex, which sends signals to the muscles to initiate movement. According to Pascual-Leone, when we imagine the movements necessary to play a piece, the prefrontal and supplementary motor areas (planning areas), basal ganglia, and cerebellum are activated. Only the primary motor cortex is not.

Since that time, studies with pianists, string players, trombonists, jazz students, and others have all shown that motor imagery has almost the same effect in the brain as actual physical practice, and the combination of mental and physical practice leads to greater performance improvement than physical practice alone.[31] Skill level at an instrument doesn't necessarily mean one will be better at motor imagery. Imagery requires practice. But the more technique and musicianship are at one's disposal, the more tools one has available to use in motor imagery. This is not to say that beginners aren't able to use motor imagery. Even a child can imagine how it feels to play a five-note tune. As with many other things, it can be made into a game with children—they love practicing away from the instrument by imagining it.

It's quite possible to imagine a variety of motor movements that don't involve sound. One can imagine climbing a ladder or walking along the street. But motor imagery for music cannot exist without auditory imagery as well. Movement is used to create sound; the two are inextricably linked.

Not only is imagery difficult to research, but it is also difficult to teach because the researcher or the teacher cannot know what the subject or student is thinking. The efficacy or legitimacy is in the results. Researchers have ways to design studies that will reliably prove, if the same brain areas are activated, that the subject must have been imagining the particular task. For a musician, whether someone is using motor imagery shows up in the results—can the piece be played or sung after spending time using imagery? Several years ago, I worked with a neuroscience major who was an extremely musical pianist. Over two summers, she had research internships at Harvard. There was no piano available to her except in a student lounge, not the best place to try to practice. The second summer we decided she would try using motor imagery to learn some repertoire, and she chose Gershwin. All summer, she practiced using only motor imagery. When she returned to campus, we had agreed that she would spend only ten minutes at the piano before her first lesson. When she played the Gershwin for me, the music was all there, right notes and rhythms, emotional expression, dynamics, phrasing. It was no different than had she physically practiced the piece. She wouldn't have been able to learn Debussy études because they were beyond her technical capabilities. But having the necessary technique to be able to play a particular piece should mean that it can be learned using motor imagery.

Imagine playing the Satie excerpt in Figure 8.1. If you play the piano, imagine playing it as written. Or imagine a line you could play on your instrument, the melody if you are a singer. Effective motor imagery involves embodied cognition. What does your body have to do to make the crescendo? What is the difference in movement if you are playing *piano* or *forte*? How does your body/arm/hand feel moving from one note or chord to the next? Can you imagine what fingerings or bowings you are using? Can you imagine different articulations if you are a wind or brass player? Are there places where you change your embouchure? Where in the phrase will you breathe? *[Lea Pearson's article "Navigating Embodied Practicing" includes a section on imagined practice. It can be found as item 8.2 on the companion website ▶.]*

Motor imagery is valuable not only for practicing away from an instrument. It is also an invaluable technique to use if injured. In addition to being

Figure 8.1 Erik Satie, *Gymnopédie*, No. 1

able to learn the motor movements of a piece through imagery, researchers have found that mental imagery exercises help preserve arm strength when an arm is immobilized. In a study in which the subjects wore a rigid cast extending from just below the elbow past the fingers, the half who regularly imagined they were intensely contracting their wrist for five seconds and then resting for five seconds lost 50 percent less strength than those who did not do the imagery exercises. That can be quite significant for someone immobilized for several weeks who doesn't want to lose strength.[32]

A note of caution, however: focal dystonia and other movement disorders are often caused by playing with too much tension or with faulty hand or embouchure position. Imagining that same amount of tension or same faulty position when engaged in motor imagery will exacerbate the problem, not help. Movements in the mind must be without any stress or tension. Fortunately, stress-free and tension-less movement is easier to imagine than it can be to physically achieve.

Back to Fei-Ping . . .

When Fei-Ping was listening to Rachmaninoff and Horowitz and imagining himself playing, or when he imagined playing as he was picking rice kernels or carrying heavy equipment, he was, in fact, practicing. All areas of the brain involved in making music were activated, with the exception of the motor cortex. Motor imagery, unfortunately, did not prevent Fei-Ping from developing a hand injury due to the forced heavy labor as well as the stress of performing without practice. But medical treatment after the Cultural Revolution addressed the hand injury, and the ten years of practicing in his mind enabled him to again return to playing the piano and developing a career. His experience is unique, but practicing motor imagery can be helpful to anyone who plays a musical instrument. Practice is with the brain, not the muscles. If you can play a piece in your mind using motor imagery, you will be able to play it at your instrument.

After the year spent at the Eastman School, Fei-Ping, shown in Figure 8.2, transferred to Juilliard. He lived on 92nd Street on the Upper West Side in New York with his brother, Fei-hsing Hsu. During the early 1980s, he was entering and winning several competitions. He was named the 1983 gold medal winner in the Arthur Rubinstein International Piano Master Competition, received second place in the 1983 Gina Bachauer International Piano Competition, and was named the prizewinner for best performance of the commissioned contemporary piece in the 1982 University of Maryland International Competition (later named the William Kapell International Piano Competition). The competition commissioned a new work each year and the pianists had three months to learn it. The commissioned work in 1982 was Richard Faith's Fantasy No. 2.

His brother expressed surprise that Fei-Ping had won the prize for the best performance of the contemporary work because he had never heard him practice it. He didn't even know it was in the competition. Fei-Ping's response: "I practiced playing it in my mind every day when I walked from our apartment to Juilliard." That was about a thirty-minute walk, and that daily walk was the only time during which he practiced the piece. Fei-Ping made his debut at Alice Tully Hall in 1983 and performed throughout the United States, Europe, and China for several years before he was tragically killed in a car crash in China in 2001 while on tour.

Motor imagery made it possible for Fei-Ping to continue developing his musical skills during the ten years that he was deprived of regular practice. But it was his personal experiences and attitudes about life that influenced his emotional connection to the music and his ability to communicate that emotion to an audience. His niece Hsing-ay Hsu, herself a Steinway artist and winner of the silver medal in the 1996 William Kapell International Piano Competition, recalls studying with him during her four years in high school. He impressed on her that playing the piano was a very personal act,

Figure 8.2 Fei-Ping Hsu, pianist
Credit: Photo used by permission of the Hsu family

that you must put every bit of your emotional experience into the music, that music was an "expression of life." *[Links to a clip of Fei-Ping Hsu performing the Chopin Nocturne, Op. 62, No. 1 can be heard on the companion website at item 8.3 ⊙.]*

Key Concepts

- Musicians tend to know about, and use, visual and auditory imagery but are less familiar with motor imagery.
- Kinesthesia, our sixth sense, delivers information to the brain about effort, movement, position, and weight delivery; it is the sense used when practicing with motor imagery.
- We can learn about the power of motor imagery through the story of Chinese pianist Fei-Ping Hsu, who was able to return to playing and

develop an international reputation after not being allowed to practice for ten years during the Chinese Revolution.

- Terminology matters: the generic term "mental practice" refers to far too many kinds of mental processes. Visual, auditory, and motor imagery refer to specific brain processes that are helpful in practicing.
- Visual perception and imagery share processing areas in the brain, as do auditory perception and imagery.
- In motor imagery, all the areas of the brain involved in music processing are active, with the single exception of the motor cortex which sends signals to the muscles to move.
- Motor imagery has been shown in multiple studies to be the only kind of imagery that causes neuroplastic changes in the brain in a nearly identical way to physical practice.

9

Seeing Sound, Hearing Movement— Music and Mirror Neurons

> It is unlikely that any other artist, excepting only Paganini, has the power to lift, carry and deposit an audience in such high degree. . . . In a matter of seconds we have been exposed to tenderness, daring, fragrance and madness. The instrument glows and sparkles under the hands of its master. . . . It simply has to be heard—and seen. If Liszt were to play behind the scenes a considerable portion of poetry would be lost.
>
> —Robert Schumann, review of a Franz
> Liszt concert, Dresden, 1840[*]

The composer Robert Schumann didn't live at a time when he could hear disembodied music as one does today, in cars, living rooms, through earbuds, in elevators and shopping malls. The only way he could experience music was live, seeing the performer's movements as he listened. Yet he understood that the visual—seeing the performer—was an important part of the experience. Today, far more time is devoted to listening to music than watching live performances. We have become accustomed to music disconnected from the actual performance and don't think about the human movement that is creating the sound. Yet, when people do attend live performances, they intuitively know there is something about the experience that reaches them in a different way. There is a different kind of excitement and energy.

Several years ago, I heard the St. Paul Chamber Orchestra in a concert that included the Beethoven "Eroica" Symphony. I went to the concert wishing they were playing something else because I had heard the "Eroica" so many times. But that evening's performance was a completely different experience. The thirty-six musicians played without conductor, and the performance was electrifying. They took slightly faster tempos than usual—playing with a

The Musical Brain. Lois Svard, Oxford University Press. © Oxford University Press 2023.
DOI: 10.1093/oso/9780197584170.003.0009

great deal of flexibility and with such clarity that I heard interior voices I was not accustomed to hearing.

I usually like to sit in the balcony, but that night I was sitting near the front of the orchestra section, so I had a close-up view of the performers, their facial expressions and body language. It was like watching musicians in a string quartet: lots of eye contact, slight head or body cues among the players, an emotional intensity often missing when an ensemble has played a particular work countless times.

Power of the visual

Chia-Jung Tsay is a pianist who made her Carnegie Hall debut at the age of sixteen and earned degrees from the Juilliard School and Peabody Institute. At a time in her life when she was entering multiple piano competitions, she noticed that how she fared in competitions varied depending on whether the elimination rounds involved submitted audio recordings or live performances.

Later, as a Harvard PhD student in organizational behavior and psychology (she also received a PhD in music from Harvard), she decided to test her intuition about the impact of the visual on the decision-making process, whether the decisions had to do with how she fared in her piano competitions, decisions people made about others, or consumer decisions. Since her curiosity stemmed from her own experience in music competitions, she designed a study using high-profile music competitions. Over a series of seven experiments, 1,164 participants were given excerpts from the top three finalists in each of ten prestigious international competitions. The excerpts were in three formats: sound only, video only, or sound and video together. Each participant was given one of the three formats and asked to identify the winners of the competitions. Results were then compared to the actual winners selected by the competition judges.[1]

The first group of participants in the study were novices who had no musical training. Tsay later repeated the study with professionals who had either performed in or judged an international competition, or both. She gave the participants a pre-study questionnaire that asked what the most important factor would be in determining the winner. Both novices and professionals overwhelmingly indicated that they would make their decisions based on sound—not surprising, since these were music competitions.

But when the results were tallied, Tsay was in for a surprise. Both novices and professionals who listened to the sound-only excerpts picked the winners of the actual competition at a rate *less than the rate of chance*. Had they randomly chosen, they would have done better. On the other hand, the participants who received the visual-only component *scored significantly above chance* in identifying the winners. That result might have been expected with novices who had no musical training, but it was a shock with the professional participants. This astonishing finding suggests that, not only does the visual have an impact on our perception of the performance, but visual information overrides auditory information. Tsay conducted later studies with chamber ensembles and orchestras. The results were the same. Visual information appeared to take precedence over auditory information.[2]

The reaction of instrumentalists when learning of Tsay's research is to say this is their worst nightmare. "We spend our lives concentrating on sound, only to find that people are paying more attention to how we look. Members of the audience are watching and not really listening." Singers, on the other hand, say, "of course! We know how important the visual is because we're trained to act out the music. We have a text that is about something, and when we're singing opera, we're acting. Even when singing lieder, we use body movement and facial expressions to help convey the meaning of the song."

Since the participants who had the silent video excerpts matched the decisions of the competition judges, it would seem that the professional musicians judging the competitions were also influenced by the visual in the performances. Although individual jurors at a competition reach different conclusions about contestants, and the winner is the contestant who receives the highest rating overall, it remains the case that the individual selected by the various competition juries was also the individual most often selected by the participants in the study who saw the visual-only format.

The participants who watched silent videos were asked what visual cues led to their choices. They cited the performer's passion, intensity, movement, energy, involvement, enjoyment, confidence, and style, among other descriptors, some of the same things I had noticed in the St. Paul Chamber Orchestra performance. At this level of competition performance, everyone plays extremely well, so the differences in technical expertise are minimal. But what varies is how movement is used to create sound, how the body is used by performers to generate their technique, how involved the performer is in the music, and how intensely the performer communicates. These are all conveyed more clearly by sight than sound.

Tsay suggests that emotional contagion may play a role in the results of her studies. Emotional contagion is the tendency to have one's emotions triggered by those around you. People surrounded by depressed people much of the time are likely to become depressed. On the other hand, crowd noise will create contagious excitement for the winning team at a football game. Tsay suggests that perhaps emotions exhibited visually through a performer's body language in a live-music situation come to be shared unconsciously by members of the audience.

How do we come to share the emotions of performers we see? Why are we influenced, as Schumann was by Liszt's visual presence, or as the participants in the Tsay study were by the facial expressions and body language of the performers in the video clips? The answer may lie in mirror neurons, one of the most exciting discoveries in neuroscience of the past thirty years. These amazing brain cells provide the neurological basis for sharing the emotions or understanding the body language of others, and they provide the neurological mechanism for emotional contagion.[3]

What are mirror neurons?

The human is indissolubly linked with imitation: a human being only becomes human at all by imitating other human beings.

—philosopher Theodor Adorno[4]

Mirror neurons are fascinating brain cells that allow us to understand the actions, intentions, and emotions of others through unconscious "mirroring" or imitation in our own brains. These neurons are active when we are learning an instrument and imitating a teacher, they are active when hearing music and imitating the sound. But perhaps most surprising, these mirror neurons are active when we are listening to a live performance, and they influence how the performance is *heard*, as the Tsay study demonstrated.

We experience the world through our senses. Perception of sound happens in the auditory cortex, sight in the visual cortex, and touch in the somatosensory cortex. Acting on the information we have perceived through our senses requires neurons in the motor cortex to initiate movement. Neurons in the motor cortex do not interpret information from our senses, and neurons in the auditory and visual cortices do not plan and make motor movements.

They operate independently. At least researchers thought they did until the discovery of mirror neurons.

The discovery of mirror neurons showed that a motor neuron could have a dual function—it could function as a motor neuron *and* a sensory neuron. It could be an action neuron as well as a perception neuron. Mirror neurons are a small subset of motor neurons, perhaps 20 percent.[5] But they are activated when we perform an action ourselves, when we see someone else perform that same action, when we hear the sound of the action, and even when we see the word or phrase describing the action.

When we see a friend in tears, when we see a couple embrace on television or at the opera, when we see someone erupt in anger, when we see someone passionately involved in making music onstage, we don't react just to the visual information and cognitively process it. Some of the cells that are activated in our brain are the same cells that are activated when we feel any of those emotions ourselves.

Mirror neurons are imitation neurons, although we obviously don't carry out every action or express every emotion that we see. There are other brain cells that prevent us from imitating everything, but unconscious imitation by mirror neurons is a primary catalyst for learning. They are active within minutes of a child being born. In a now-classic experiment conducted in the mid-1970s by Andrew Meltzoff, infants between twelve and twenty-one days of age were shown on videotape to be imitating facial gestures.[6] This was before the discovery of mirror neurons but is a perfect demonstration of them.

Toddlers learn by imitation. There is no thought process involved, just imitation—of each other and of parents. As children mature, they begin to incorporate deliberative cognitive processes into learning, but mirror neurons are still being activated every time an action is observed, and learning is still facilitated by imitation through the mirror neuron network, whether it involves learning a musical instrument, learning to play tennis, or learning baking techniques with mom, as seen in Figure 9.1.

The discovery of mirror neurons

In the late 1980s, a research team in Parma, Italy, led by neurophysiologist Giacomo Rizzolatti was studying macaque monkeys to determine the brain networks involved in grasping and manipulating objects such as food. The brains of monkeys are smaller than those of humans, but similar in structure,

Figure 9.1 Young girl mirroring mother
Credit: iStock.com/aabyss

so researchers often use monkeys when initially researching something they will later explore in humans. The researchers were interested in understanding more about motor control of the hand. The hands are used more than any other part of the body, and losing hand function as a result of stroke or traumatic brain injury is devastating. If they could learn more about the brain mechanisms behind the use of the hand in monkeys, that knowledge could potentially lead to an understanding of how humans who have lost hand function due to certain kinds of brain damage might recover the use of the non-functioning hand.

Electrodes were inserted into the premotor cortex of the macaques, the part of the brain that is known for planning and carrying out movements. When the monkey picked up an object, neurons in the brain would fire, the sound was amplified, and researchers would hear it over a monitor in the lab. One day during a break in experiments, one of the researchers picked up a peanut (various stories differ on whether it was a peanut or an ice cream cone), and he heard a burst of sound from the monitor, signaling activity in the monkey's brain. But the monkey was sitting quietly, doing nothing—he had only seen the researcher pick up the peanut. The monkey had not picked up anything himself.

The research team didn't realize the significance of this burst of brain activity at the time, but after a few years and dozens of experiments in the Parma lab and elsewhere, what had happened on that day, and continued to happen in further experiments, became clear. There was a subset of cells in the premotor cortex and parietal area of the monkey's brain that linked action and perception.

These neurons fired when the monkey picked up a peanut, but also when he saw someone else pick up a peanut. A different set of neurons fired when the monkey broke open the peanut or saw someone else break open the peanut, and other neurons fired when the monkey put the peanut in his mouth or saw the experimenter do so. Some responded to the sound of the peanut shell being broken.[7] Some of these neurons were found to code for intentions. When a peanut is picked up, is it intended to be put in the mouth or placed on the table?[8] Different cells fired for different actions, and a system had been found in the monkey's frontal and parietal lobes that made it possible to understand actions and intentions.[9]

The neurons were initially referred to as "monkey see, monkey do" neurons, but eventually named mirror neurons. Because the purpose of the original studies had been to understand hand actions in monkeys as a way to potentially learn more about human hand actions, researchers began to look for similar neurons in the human brain.

Mirror neuron systems in humans

Electrodes implanted into a macaque's brain could detect a single mirror neuron, but most humans are not willing to have electrodes implanted in their brains for the sake of research. So in the first study of humans, Rizzolatti and colleagues used transcranial magnetic stimulation (TMS), the same technology that was used in the first study of neuroplasticity in adult humans (see Chapter 4).

They measured human subjects while they were observing a researcher looking at objects, grasping those objects, and tracing geometrical figures in the air with his arm. TMS could not detect activity at the level of a single cell in human brains, but it did detect a network of neurons that were in the same brain areas as the individual mirror neurons in macaques and seemed to serve the same function. They called this a mirror neuron system (MNS).[10]

Later studies using magnetoencephalography (MEG) or functional magnetic resonance imaging (fMRI) confirmed that humans have a mirror neuron system in the premotor cortex. It is active when performing actions as well as when observing them, when hearing the actions or when hearing those actions described, and it is involved in imitation. And as with monkeys, the mirror neuron system in humans is instrumental in understanding intentions. Based on the context of the situation, is someone about to begin a bicycle ride, or has he just finished?[11]

Research expanded into other areas, and emotions were explored in connection with the mirror neuron system.[12] Researchers found that the mirror neuron system was connected to the limbic system, which is instrumental for emotional processing and behavior. Although we think of emotions as internal, they are communicated by the voice as well as by facial expressions and body movements that can be seen, and these are controlled by motor processes. Our brains have the motor templates for emotions and when we see emotions in others, whether on a film screen or watching a performer at a music concert, we understand those emotions through our mirror neuron system.

Other studies followed, confirming that these mirror mechanisms are at the heart of understanding the emotions of others. Some referred to this link of emotions between the observer and the observed as *embodied simulation*.[13] Embodied cognition, as earlier described, refers to understanding through the body rather than through the mind. The way one moves one's body, the way one sits or stands, and what one touches can all influence how one thinks about a situation. Embodied musical cognition refers to feeling and understanding music through one's body—moving to a beat, for example. *Embodied simulation* means that by way of the mirror neuron system the observer is experiencing the same way of knowing through the body as the person being watched. When Dr. Tsay suggests that embodied cognition may be behind the results of her study, mirror neuron researchers would suggest that mirror neurons are the basis of embodied cognition/simulation.

The theater world immediately understood the significance of mirror neurons, and a great deal has been written about mirror neurons and acting. In a story attributed to an Israeli actor, the actor noted that "while neuroscientists found this property extraordinary, they should have asked 'us actors,' who have known—or better, 'felt'—all along that they must have something like these cells in their brains. When I see someone with a painful facial expression, said the actor, I feel her pain inside me."[14] Peter Brook, the legendary English film and theater director, has spoken often about mirror neurons and their importance in the theater world, commenting that neuroscience has provided a biological explanation for the sharing between actor and spectator, and that is the basis on which the theater evolves—and revolves.[15]

Mirror neurons were quickly embraced by the scientific community and researchers began to think that these cells might play a role in any number of areas. Research expanded into the role of mirror neurons in empathy, learning, the evolution of language, the neural foundation of autism, and theory of mind (recognizing one's own mental states and realizing that other people have mental states different from one's own). Not surprisingly with a scientific discovery that claims to do so much, there has been somewhat of a backlash. For many scientists, one of the problems with mirror neurons was the leap made to humans from the original experiments with monkeys. Some scientists questioned how it was possible that a cell in the motor cortex of a monkey could provide a neural basis in humans for language, empathy, autism, and more.

But in 2009 for the first time, single-cell mirror neurons were also detected in humans. A team at UCLA was able to piggyback onto surgical procedures for seizures being performed on twenty-one patients with epilepsy (with the patients' consent). The patients had already been implanted with electrodes to identify areas for surgical treatment, somewhat in the manner of the early work by Wilder Penfield (see Chapter 4). The placement of the electrodes was based on the medical data of each particular patient, so the location of the electrodes was not necessarily in the areas known for mirror neuron activity in monkeys.

Researchers found that there were indeed individual neurons that fired both when a patient performed a task and when he observed the same task. Interestingly, the mirror neuron activity was found in two areas that had not previously been recorded, either as single-cell activity in monkeys or systems activity in humans. These areas were associated with movement selection and with memory.[16]

This study provided evidence that mirror neurons exist in humans, and because this activity was found in areas that had not previously been associated with mirror neurons either in monkeys or in humans, researchers suggest that mirror neurons may be found in more areas of the brain than previously thought. While there may be some disagreement about the role of mirror neurons in autism, language, or theory of mind, there is no disagreement among researchers about the important role of mirror neurons in observation and imitation.

Studies of mirror neurons in musicians

Many musicians discover their fingers moving when listening to a piece that they play, as though they are themselves playing. In the late 1990s, researchers decided to explore what was happening in the motor cortex of pianists listening to music they knew but were not playing. Is there something in the brain that connects our perception and our production of music? German researchers designed an experiment using magnetoencephalography (MEG) to compare the motor cortex in pianists and non-pianists while they listened to piano pieces.[17] They found activity in the motor cortex of pianists while listening, but not in the motor cortex of non-pianists—a connection between listening and motor activity (see Chapter 4).

Another study asked professional and amateur violinists to tap out the first sixteen bars of Mozart's G Major Violin Concerto—without producing any sound.[18] The professional violinists showed significant activity in the auditory cortex even though there was no sound, the amateurs did not. These studies showed a connection between auditory and motor areas in the musician's brain that goes both directions—motor to auditory and auditory to motor.

At about the same time, researchers also explored a pianist's brain activity when just watching someone play the piano. Using fMRI, researchers compared professional pianists to non-musicians while they observed piano playing without sound. The pianists showed significant activation in their motor cortex watching the piano being played, while there was less activation in the motor cortex of non-musicians. There was also activation in the auditory area of the brain of the musicians, even without sound, but not the non-musicians. The researchers suggested a musical "language" linking visual, auditory, and motor areas of the brain acquired by observation and imitation.[19]

These early music studies weren't studying mirror neurons; they had other objectives. But they showed motor activity in the brain as a consequence of listening to or observing music-making, and that's what mirror neurons have been shown to do. Researchers began to recognize an extended neuronal network in musicians that corresponded to a mirror neuron network.[20]

What do mirror neurons mean for teaching and learning?

When watching someone perform a motor activity, whether sports, playing a musical instrument, or dancing, mirror neurons in the brain are activated. If it is an activity in which we ourselves engage, our mirror neurons are strongly activated. Even if we do not engage in that activity, mirror neurons will still be activated to some extent because we are mirroring general movement patterns.

Observing

The great pianist Vladimir Horowitz referred to imitation as caricature. Perhaps it is, but can you imagine learning to play tennis if you have never seen anyone play? Imagine a coach giving you verbal instruction only—telling you how to hold the racket, explaining a serve, describing a backhand. The difficulties in learning would be enormous. On the other hand, watching someone's backhand or serve facilitates the learning process because of the imitation system in the brain that maps what we see onto our motor system. Yes, cognition also plays a role. We analyze why we are making a certain movement, and what we need to do to make it more efficiently. But when we observe motor activity that we currently cannot do that has some relationship to a move we are able to do, we are primed to be able to learn the more complicated skill because our mirror neuron system provides a template.

Any new motor pattern is learned by rearranging already known elementary motor acts into a new pattern. A beginning piano student learns the most basic patterns of hand position by watching the teacher and practicing, and because imitation happens so readily, it's extremely important that the teacher model a healthy technique. As the student observes the teacher or someone else play more complicated patterns, that visual signal is processed and mapped onto the motor counterpart by the mirror neuron system—building on the actions the student already knows how to do.[21] That doesn't mean the more complicated movements will instantly be learned. That's the purpose of practice, but mirror neurons provide a template for learning the more complicated pattern.

A figure skater learns to do a salchow jump before a toe loop, a lutz before an axel. A pianist must learn basic five-finger patterns before playing a scale, but watching someone demonstrate how the thumb goes under the

Figure 9.2 Young pianist mirroring hand position
Credit: Dana Olsen

palm when playing a scale makes it much easier than having a teacher simply describe how to do it. (See a young pianist mirroring in Figure 9.2.) A beginning violinist must learn how to hold the bow before learning any bowing techniques, and both are easier to learn if someone is demonstrating. There is a progression in learning, and we always have to build on what we know. What wasn't clear until the discovery of mirror neurons is the importance of observation in learning new techniques.

Listening

Mirror neurons are also auditory, so what about listening? I studied with a wonderful pianist in undergraduate school, and he taught me a great deal about communication and musicality. But he was adamant about not listening to recordings of pieces that I was working on for a performance. He didn't want me to copy someone else's interpretation but to develop my own. I understand his reasoning, but had he known about mirror neurons, he may have thought differently.

Suppose you are learning a Beethoven piano sonata or an Ives violin sonata. You know the piece quite well and have a fairly clear idea about the interpretation you want. You listen to two or three recordings of the work by pianists or violinists you admire, and you also hear a live performance. In those three or four performances, you will hear a variety of interpretations, each one involving different motor execution. Tempi will be different in places, *ritardandi* and *accelerandi* will vary, dynamics and phrasing will differ. Because of mirror neurons, the motor areas of your brain are activated, as though you yourself are playing with each different interpretation you hear.

Cognitively, you will think about the different interpretations, perhaps try some ideas and discard others. But strictly on an unconscious level, the motor areas of your brain have been activated and have planned the movements necessary to create the different interpretations that you have heard. Does that mean you would instantly be able to imitate someone else's interpretation? Of course not, that takes practice. But because your motor areas have been activated in response to hearing different interpretations, you will be primed to learn them, and you can build on that should you choose to do so. Many teachers direct their students to listen to several recordings of pieces they are working on. It is important that students be directed to quality performances because they may not know what performers to listen to and they will be mirroring whatever they hear—musical or not.

Singers often say that if they listen to a singer with tension in the voice performing repertoire they themselves sing, they experience tension in their own vocal folds. They are mirroring what they hear. Researchers refer to this as subvocalization in singers and suggest that this may point to vocal resonance or empathy in a singer's perception. But it also points to mirror neurons.[22]

Several years ago, I met a sixteen-year-old student named Sue when she began her studies at a major music school. She had lots of fingers, as pianists say, but little sense of style. But by the time she was finishing her four-year degree, members of the piano faculty all spoke about her amazing sense of performance style and how wonderfully she communicated the essence of the music. She practiced several hours each day, but she also attended virtually every solo, chamber, and orchestral performance during the four years she was in residence. If there was an admission fee, she would usher so she could attend for free. By the end of four years, she had heard an amazing amount of repertoire in a variety of styles, interpretations, and genres. She also listened to a lot of recordings.

Of course, she was studying with a piano professor who would have been giving her extensive coaching. She no doubt discussed the concerts with her friends, compared performances, spoke about what worked and what didn't. She would have thought a great deal about the music she was hearing and done a lot of cognitive processing. But her auditory and audiovisual mirror neuron systems were also constantly being activated by what she heard and saw—activated strongly if she heard music she herself was playing but also activated to some extent listening to music she didn't know, because many of the sound patterns and movement patterns were already familiar to her from studying other music. She became a spectacular pianist in part due to unconscious learning through her mirror neuron system.

Musicians, mirror neurons, body language, and emotions

Music has been regarded for centuries as the "language of emotions," but whose emotions? The composer's, the performer's, or the listener's? There is no direct line from what the composer may intend to what the listener hears. A composer manipulates elements of music such as pitch, rhythm, articulation, dynamics, timbre, and choice of instrument or ensemble to convey emotion. He conveys his intentions through a musical score, an approximation at best of what he is hearing in his own mind. The performer recreates and interprets that score in a way that may or may not be in line with the composer's intentions, and the resulting music reaches the listener, who brings his own emotional mindset to the music being performed.

Some composers have no intent to convey emotion. In speaking about expression in music, Igor Stravinsky said, "I consider that music is, by its very nature, essentially powerless to express anything at all."[23] Yet one cannot listen to Stravinsky's *The Rite of Spring* or *The Symphony of Psalms* and not feel some kind of emotion, whether Stravinsky intended it or not.

Whether one is watching and hearing a conductor and orchestra perform *The Rite of Spring*, a pianist perform a Chopin ballade, or a violinist perform a Bach partita, one sees the body movements of the performers creating the sound. So as researchers had done with initial mirror neuron research, they began to look at the role of mirror neurons in conveying emotion in musicians. They looked at the relationship between performers' body movements and facial expressions on listeners' perception of emotion and at the role of the visual in coordination in ensembles. One study after another confirmed a mirror neuron system involvement in conveying emotion in music.[24]

Studies discussed earlier in this chapter showed that the connection of the mirror neuron system and limbic system was at the heart of understanding emotions in others. This was also confirmed by researchers studying how emotion is conveyed in music. They concluded that this connection between the mirror neuron system and the limbic system allows people to understand complex musical signals and their emotional response to them.[25] One study concluded that "Imitation, synchronization, and shared experience may be key aspects of human musical behavior."[26] Listeners share the experience, including the emotions, with the performers they are hearing. The same neurons are firing whether watching a performance or performing.

Actors embrace the idea of mirror neurons; they understand the power of the visual. Singers do also, but most instrumentalists really haven't given it much thought, and are concentrating on communicating the sound. But if, as Peter Brook suggests, mirror neurons are a biological explanation for the sharing between actor and spectator, then they are also the biological explanation for the communication between performer and audience, between teacher and student, between conductor and orchestra, and among players in an ensemble.

What you see is what you hear

Music is the emotional life of most people.

—Leonard Cohen

Several years ago, I read a review of concerts by two pianists that provided a perfect illustration of mirror neurons. The concerts were held at the International Keyboard Institute and Festival, then at Mannes College in New York City, now held at Hunter College. One pianist was a young Russian who had just won the Franz Liszt Competition. The other was a more seasoned American, older by a few decades. They played similar repertoire, including Liszt and either Debussy or Ravel.

Anthony Tommasini, music critic for the *New York Times*, wrote about the young Russian's prodigious technique, myriad shadings, and demonic fervor, but concluded that he appeared to be miserable, and didn't seem to enjoy playing the piano. In contrast, he spoke of the older American's avuncular charm, his eagerness to talk about the music. He wrote that the American lacked some virtuosic dazzle and sonic power and mangled some octaves, and may not have technique to burn, but Tommasini ends the review with this statement: "Still, he played all three works with musical authority and pianistic flair. During each performance I kept thinking about how astonishing these pieces are. If a pianist can convey this, he is a master in the ways that matter most."[27]

What matters most is communication. The performances weren't simply about technique, they were about communicating to the audience the emotional content of the music. Musical communication—emotion—is conveyed not only through the music itself, but through the facial expressions and body movements of the performer. Gestures and movement reflect the physicality that produces the sound, tone colors, dynamics, expression, and sometimes the harmonic movement of the music itself. That physicality is better conveyed through the visual than through sound.

It is relatively recent in our evolutionary history that we have divided ourselves into listeners and performers. In the more distant past, music was a social activity, and everyone made music. This is still the case in many cultures today. Whether drumming, singing, or playing a flute, making music involves movement. The mirror neuron system provides for a sharing of that motor activity, of that musical communication. I feel your movement and therefore your emotions because of mirror neurons.[28]

Nadia Boulanger, the great French composer, conductor, and teacher, once said, "Music can never be more or less than who you are as a human being." Emotions, passion about making music, and commitment to communicating what is in the score are internal feelings, but they are seen externally through facial expressions and body language and mirrored by others. How musicians use their bodies to create the sound and expressiveness they want is also a reflection of who they are as people and as musicians, and that is mirrored by others.

Or as American bandleader Ray Conniff put it, "If you believe in your art and you love what you do, that energy will go out and people will respond." That energy is communicated through mirror neurons, and these miraculous little cells are what make it possible for us not only to learn a musical instrument, but to enjoy the communicative power of music.

Key Concepts

- Mirror neurons, brain cells that are activated for both action and perception, may be one of the most important discoveries in neuroscience in the past thirty years.

- People mirror others, and are mirrored themselves, whether aware of it or not.
- Because imitation via mirror neurons is a primary way of learning, visual and auditory demonstration and imitation are crucial for teaching and learning.
- When teaching any motor skill, whether tennis or the piano, model a healthy technique because what you do will be mirrored.
- If teaching music, direct students to high-quality performances for listening; their mirror neurons will mirror what they hear.
- Be aware of the power of the visual when listening to live performances, whether classical, rock, or jazz. Are you responding more to the visual than to the sound?
- When you play your instrument, generate the sound in the most healthy, authentic way possible because your listeners are simulating—mirroring—your movement and your emotions.

10

Does Music Really Make You Smarter?

The theory of relativity occurred to me by intuition,
and music is the driving force behind this intuition.
My new discovery is the result of musical perception.
                    —Albert Einstein, 1921 Nobel Prize in Physics*

Many people in politics, science, medicine, and other professions have attributed their success to their study of music. What is it about music that promotes success in other areas, or is that just a myth? The question has been raging for nearly thirty years, precipitated quite unintentionally by a young researcher in California. In October 1993, Frances Rauscher and colleagues published a single-page paper in the journal *Nature* titled "Music and Spatial Task Performance."[1] Rauscher was a psychologist at the University of California at Irvine and had been a cello performance major in college. She was interested in research going back to the 1970s that suggested a connection between music, mathematics, and spatial reasoning. In the *Nature* paper, she described a study in which thirty-six college students were given three sets of standard IQ spatial-temporal reasoning tasks after ten minutes of listening either to Mozart's Sonata for two pianos in D major, K488, listening to a relaxation tape, or silence.

Rauscher found that the students' spatial IQ scores were eight to nine points higher after listening to Mozart as opposed to either the relaxation tape or silence. Rauscher made it very clear that (1) the effect didn't last for more than ten to fifteen minutes; (2) the results applied *only* to spatial-temporal reasoning skills; (3) effects were unknown with other composers since Mozart was the only composer used; and (4) she suggested that listening times could be varied to see if any other measures of general intelligence, such as verbal reasoning, quantitative reasoning, or short-term

The Musical Brain. Lois Svard, Oxford University Press. © Oxford University Press 2023.
DOI: 10.1093/oso/9780197584170.003.0010

memory, might also be facilitated. The study *made absolutely no claims about general IQ* because it wasn't measured.

Rauscher thought that the findings in this study were rather "neat," but didn't think anyone else would be particularly interested. Much to her surprise, the Associated Press called her before she even knew the paper had been published. News outlets from the *New York Times* to the BBC picked up the story with headlines that proclaimed, "Listening to Mozart Makes You Smarter," or "Music Makes You Smart." An enterprising journalist named it the "Mozart Effect," and the public was hooked. The idea that one could derive some cognitive benefit just by listening to the music of a particular composer was intriguing. The term has been in use ever since to suggest that listening to Mozart, or classical music in general, makes you smarter, something that Rauscher never suggested or even contemplated.

In the nearly thirty years since Rauscher's study was published, thousands of research studies have looked at the effect that studying music has on cognitive development. While there is no evidence whatsoever that listening to music makes you smarter, a great deal of evidence has accumulated that learning to play a musical instrument and making music gives you a lifelong cognitive advantage in many other areas. The idea that training or developing skills in one area or domain might have an effect on other domains is called *transfer*. *Near transfer* occurs when there is a close relationship between the training domain and the transfer domain. For example, people who have studied a musical instrument are better at pitch and rhythm discrimination and have faster finger-sequencing abilities (tapping sequences such as 51324 or 14235), skills closely associated with playing an instrument.

In *far transfer* effects, there is a transfer of skills, knowledge, or cognitive processes gained from one domain to another that has no apparent connection. Researchers have demonstrated the positive effects of musical training in the areas of language, math, speech perception and processing, memory, emotional control, executive function, and more—areas that do not have a close relationship to music. This is far transfer. Many of the scientists doing work in these areas argue enthusiastically for the inclusion of music in the K–12 curriculum because they see firsthand what learning an instrument can do for a child—or an adult. Three areas of far transfer are particularly interesting: the impact of the study of music on math, on executive functions, and on speech processing and language.

Music and math

> Just as music comes alive in the performance of it, the same is true of mathematics. The symbols on the page have no more to do with mathematics than the notes on a page of music. They simply represent the experience.
> —mathematician Keith Devlin[2]

There has always been an interest in the relationship between music and math. Pythagoras, who died around 500 BC, is considered by some to be the founder of both math and music—famous for his theorem on triangles in mathematics, but also for developing the concept of intervals in music. There are currently thousands of books on the relationship between music and math, ranging from historical perspectives to explorations of acoustical, theoretical, physical, or analytical relationships. The American Mathematical Society has a webpage devoted to the connection between math and music, with links to several podcasts and videos in which musicians and mathematicians explain music in terms of geometry, algebra, calculus, differential equations, harmonics, fractals, the patterns of symmetry found in both, and more. Teachers have often noticed that students who are good at math are either studying or have studied music. Until somewhat recently, it has always been assumed that the correlation existed because a student who has the discipline to study music will also be disciplined in other academic areas. But many studies point to a connection other than shared discipline in studying. *[See the accompanying website, item 10.1, for a link to the "Mathematics and Music" page on the American Mathematical Society website ⊙.]*

Research funded by the NAMM Foundation (National Association of Music Merchants) found that elementary students in top-quality school music programs scored 20 percent higher on mathematics tests than children in schools without a music program.[3] A 2006 study from the Center for Arts Education Research at Columbia University tracked students in five elementary schools who were studying the violin. Over a period of six years, those students consistently showed greater improvements in standardized test scores in mathematics than did control groups at the same grade level.[4] Data from the College Board showed that students who took four years of high school arts and music classes scored an average of 92 points higher on the Critical Reading and Mathematics portions of the SAT than students who had taken only a half-year or less.[5]

Critics point out that many of the studies comparing math and music are correlational, not causal, that control groups of students who were equally matched and tested prior to the study were not used to prove that it was actually the music study that made the difference in mathematics achievement. An oft-cited report by Kenneth Elpus says that there is no link between participation in music courses in high school and higher standardized test scores, that the higher scores are due to pre-existing differences between students who elect music courses and those who do not.[6]

Martin Bergee, professor of music education at the University of Kansas, deliberately set out to disprove the link between musical and mathematical achievement. He was suspicious of the many studies that showed a relationship between music and math and thought that if you controlled for all the variables, that relationship would go away. To his surprise, the complex study of over 1,000 middle school students from seven Midwestern school districts showed the opposite.[7] He designed two models, one for the relationship of music achievement to reading achievement, and one for the relationship of music achievement to math achievement. He accounted for background variables such as grade level, gender, educational attainment of parents/guardians, socioeconomic status, ethnicity, and urbanicity. He also accounted for variables in district achievement, district behavior, available funds, and local revenue. Although he expected to disprove the possibility of correlation, he found a very strong relationship between music achievement and reading/math achievement. While most previous studies have looked at *participation* in music as related to math achievement, Bergee's study looked at *achievement* in music as related to math achievement, a significant difference because it suggests that the more skilled you become at your musical instrument, the more transfer there is to better math and reading skills.

Martin Guhn and colleagues at the University of British Columbia studied the school records of more than 112,000 students in British Columbia who had finished the last three years of high school and who had taken at least one standardized exam for math, science, or English.[8] Thirteen percent of those students had participated in at least one music course in grades ten, eleven, or twelve. The researchers found that students who participated in music had higher exam scores in all three of those areas, and the higher their achievement in music, the higher the exam grades. Again, this study looked at achievement, not just participation. In an astonishing finding, students who had begun playing a musical instrument in elementary school and were now playing in high school band or orchestra were about one academic year ahead of their peers in their English, math, and science skills, as measured by the exam grades. As in the Bergee study, Guhn controlled for gender, ethnicity, and socioeconomic background. It is difficult to refute the results when every possible variable has been considered and the study looks at progress measured over time.

Why is there a relationship between music and math?

The most basic relationships between math and music are easy to see. Elements of music, such as pitch, rhythm, tempo, form, and meter, can all be related to measurement of time and frequency, which are mathematical concepts. For example, Pythagoras discovered that musical intervals are mathematical ratios: an octave is in the frequency ratio of 2/1, the perfect fifth is 3/2, and the perfect fourth is 4/3. Meter signatures are written as fractions, such as 4/4 or 6/8 or 3/2. Whole notes can be divided into four quarter notes, eight eighth notes, etc. Rhythms are subdivisions of a basic pulse. Pitch is primarily a function of frequency, which is the number of cycles per second of a vibrating string (violin) or a column of air (flute), and all musicians are familiar with the concept of A = 440 Hz (the A above middle C on the piano corresponds to the frequency of 440 Hertz). This is all related to arithmetic, but as one mathematician put it, arithmetic is simple calculation while mathematics is cognition; arithmetic is about numbers, mathematics about theory.

At the cognitive level, both music and mathematics use symbols to represent meaning. Neither can be expressed in a written language such as English. Music notation uses symbols to give us information about pitch, rhythm, meter, and dynamics, and as we read those symbols, we translate them into sounds, either in our minds or on an instrument. We hear the representation of those symbols when we listen to or make music. Mathematics has a symbolic language of its own that expresses thoughts, facts, operations, or relationships. In both fields, symbols have meaning. *[Examples of both music and mathematics symbols can be seen on the companion website, item 10.2 ⊙.]*

Gottfried Schlaug, professor of neurology and director of the Music and Neuroimaging Laboratory at Beth Israel Deaconess Medical Center and Harvard Medical School, suggests that neuroplasticity that occurs in brain regions involved in musical processing "may have an effect on mathematical performance because of shared neural resources involved in the meaning of symbols and the mental manipulation of symbolic representation."[9] The manipulation of symbols and what they represent, whether in math or music, is called *spatial-temporal reasoning*, exactly what Frances Rauscher was measuring in her 1993 study.

Spatial-temporal reasoning

Spatial reasoning or spatial cognition refers to the ability to think about objects in three dimensions. Spatial-temporal reasoning not only requires spatial imagery or the ability to conceptualize objects in three-dimensional space, but the ability to mentally manipulate them over a period of time. People who are good at spatial-temporal reasoning (also called ST reasoning) are good at seeing how things fit together and how they can be manipulated. You are using spatial-temporal reasoning in your daily life when you merge into ongoing traffic, judging where the other vehicles are and what your speed needs to be to merge into the traffic without hitting someone. You are using spatial-temporal reasoning when you look at a map and visualize several routes to get to your destination. And you are using spatial-temporal reasoning when you fit various-sized items into a suitcase, the car trunk, or a box—without resorting to repeated trial and error. (During one long-ago move, a composer friend would look at a space in the U-Haul truck, scan the various-sized boxes or objects in the driveway and pick one that fit perfectly, as though it had been measured for that spot—great spatial-temporal ability.)

Spatial-temporal tasks on an IQ test involve such tasks as looking at an object and determining which of several choices is a rotation of that object or seeing an unfolded cube with a different design on each side and having to determine which of several versions of the folded cube corresponds with the unfolded one shown. These are the kinds of tasks the participants in the Rauscher study were asked to do after listening to Mozart, a relaxation tape, or silence. *[A link to examples of spatial-temporal tasks can be seen on the accompanying website at item 10.3 ⊙.]*

Without being aware of it, musicians use spatial-temporal reasoning all the time. Time and space are inextricably combined in music. Rhythm and melody move left to right in space on a notated page and are a function of time. Pitch, or frequency, is notated on a vertical dimension (the musical staff), is heard as low to high, and is a function of space. Patterns in music are transposed or inverted in time and space. Scales or arpeggios can be played in any key—the same pattern of whole steps and half steps, but at a different pitch, meaning higher or lower in pitch space.

One thinks about communicating the music when performing, not about the space-time relationship. But nonetheless, spatial-temporal thinking is always involved. It's there in the transposition or inversion of a Bach fugue subject. It is employed to understand meter signatures, subdivisions of a beat, the concept of a piece unfolding over a defined period of time. Composers create scores for any number of instruments, visualizing and hearing in their minds how all the parts fit together, and conductors do the same when learning a score. Music exists within time and space, and musicians learn to manipulate the spatial information concerning pitches in relationship to the time information of note values, rhythm, and meter.

Spatial-temporal processing is also crucial in STEM fields (science, technology, engineering, mathematics), in architecture, and in chess. It is important for conceptualizing solutions to the multi-step problems found in all those areas. Engineers and architects must visualize how the parts of a building or a bridge will fit together, and how each construction element builds on another. Chess players must be able to visualize several moves ahead—both their own and their opponents'. Many math educators advocate the teaching of music because of shared spatial reasoning.

Rauscher realized immediately after her 1993 study that it wasn't listening to music that would strengthen spatial-temporal reasoning but learning to play and studying a musical instrument. Since that time, many studies by Rauscher and others have shown that studying music enhances spatial-temporal reasoning. An analysis in 2000 of over two dozen studies involving children between the ages of three and fifteen who studied a variety of instruments for periods of weeks to several years and including evenly matched control groups that did not study music, showed that studying music improves spatial-temporal reasoning, a skill that is necessary for understanding mathematics.[10]

Music and executive functions

The arts train a person in discipline, independent action, thinking, and in the need for attention to detail without becoming a prisoner of that detail. . . .

I wish I could still be a bassoonist—it was a lot harder than being a scientist.

—Thomas Südhof, 2013 Nobel Prize in Physiology or Medicine[11]

A full-size orchestra is made up of 80 to 100 highly skilled individuals, each on a particular instrument, each instrument having a different role or voice within the orchestra. Someone needs to develop an interpretation of the piece, unify the tempo, shape the sound of the orchestra, make sure that players have good cues to follow so they play together, and communicate phrasing and dynamics. That is the role of the conductor. Just like individual players in an orchestra, human brains have multiple networks or areas that have different functions. The set of skills referred to as *executive functions* (EFs) serves as the brain's conductor, allowing people to set goals, plan and organize, and follow through to get things accomplished and achieve objectives.

What are EFs?

Many skills are referred to as executive function skills, but the three key areas of executive function are *inhibition* or *inhibitory control, working memory* (also called *updating*), and *cognitive flexibility* (also called *switching/shifting*). The higher-level cognitive functions of reasoning, problem-solving, and planning are built on the three core areas—as is creativity. These skills are used every day in jobs, in personal relationships, and in managing daily lives.

Inhibitory control, sometimes called self-regulation or impulse control, means the ability to control attention, focus on goals, control behavior and emotions, and prioritize and stay on task despite distractions. Good inhibitory control means choosing how to act rather than being at the mercy of emotions or impulses. A person with poor inhibitory control reacts reflexively or impulsively, striking back at everyone and everything without thinking about consequences, flitting from one idea or task to another without prioritizing and without focus.

Working memory (WM) holds information in the mind so that it can be used or mentally manipulated. It has traditionally been thought to consist of two systems, verbal and visuo-spatial. Verbal working memory is needed to make sense of a lecture or a story or even a long sentence—remembering what came first to relate it to what comes later. Nonverbal working memory is visual and refers to holding images in one's mind. Looking at a map and keeping the image in mind while you drive to a specific location involves visuospatial working memory. Working memory is important for mentally seeing the relationship between ideas, for reasoning and critical thinking.

Various arguments or ideas are kept in mind as they are sifted through and decisions are made. It is important for seeing the connections between seemingly unrelated things, which is crucial for creativity.

Cognitive flexibility is the ability to think "outside the box," to look at an argument or idea from different perspectives, to see an image differently by shifting it in your mind. It is the ability to adapt to new circumstances, such as moving to a new city for a job even if the location isn't appealing. It is the ability to accept being wrong, to take advantage of new opportunities, to adapt when situations change.

EFs have been found to be a better predictor of school readiness and success throughout school than IQ,[12] and EFs account for more than twice as much variation in final grades than IQ, even in college.[13] Adele Diamond, neuroscientist at the University of British Columbia, says EF skills are critical for success in a career, in quality of life, as well as in mental and physical health. Poor EF skills have been shown to lead to social problems, including crime, reckless behavior, emotional outbursts, and violence.[14] EFs have traditionally been associated with the prefrontal cortex, the "thinking" area of the brain, but there is growing evidence that other areas of the cortex, subcortical areas, the brainstem, and the cerebellum are also involved. EFs develop over the course of our lives and can be improved at any time—even as adults. Evidence is mounting that studying a musical instrument improves executive function skills.

How does studying a musical instrument improve EF?

A look at the three core EF areas—inhibitory control, working memory, and cognitive flexibility—shows that musicians use all three. Inhibitory control is used when deciding to practice rather than party, to focus on a section of a piece rather than just play it through, to allocate practice time in the best way to prepare for an upcoming performance. Musicians use inhibitory control when they don't insist on their own interpretation but stay open to colleagues' ideas.

The second of EF's core areas, working memory, is used when sight-reading a new score. Short-term memory holds excerpts of music in memory, but working memory is necessary for the brain to plan the movements needed to play what we just read. Working memory is used when trying different interpretations to see which we prefer or to keep in mind what a conductor or teacher has said and be able to incorporate it into our interpretation of the piece.

The third core EF, cognitive flexibility, is necessary to be able to adjust quickly if a conductor or colleague suddenly takes a faster tempo. A musician adjusts for each new hall, a pianist for each new instrument played. Ensemble performers must be flexible in taking each colleague's ideas into account and must be able to shift their focus among the voices of the different instruments. Musicians problem-solve each time they learn a new piece—what fingering to use, how to play 4 in the time of 5, how to pedal for greater clarity, how to bow a passage for best effect. A musician self-monitors—measuring current performance against how he wants it to sound.

Musicians can relate to needing every executive function skill. And sure enough, researchers have found that not only do musicians have stronger executive function than do non-musicians, but studying a musical instrument improves executive function. Becoming better at EF skills requires practice, and with music practice, EF skills are strengthened as well.

In 2014, Dr. Nadine Gaab and her research team at Boston Children's Hospital found that adults who were musicians had better working memory and enhanced performance in cognitive flexibility and verbal fluency than adult non-musicians.[15] Children between the ages of nine and twelve who had studied music for at least two years showed better verbal fluency and faster processing speeds than children who had not studied music. Processing speed refers to the pace at which one takes in, makes sense of, and then acts on information. The speed itself is not an executive function skill, but slow processing speed hinders the effectiveness of EFs. Since this was not a longitudinal study in which the children and adults were analyzed both before and after training, the question arises: might individuals with already strong EF skills have gravitated toward the study of music?

That question was answered by research conducted by James Hudziak, director of the Vermont Center for Children, Youth and Families at the University of Vermont Medical Center. He found that it wasn't a matter of individuals with strong EF skills gravitating toward music; playing a musical instrument accelerated cortical organization in EF skills. Using a database compiled by the National Institutes of Health between 2001 and 2007, Hudziak analyzed brain scans of 232 children between the ages of six and eighteen who had studied a musical instrument (average of two years). Brain scans taken over a period of years showed that not only did the cortex change in thickness in motor and auditory areas of the brain (as we saw in

Chapter 4), but studying a musical instrument also correlated with changes in the behavior-regulating areas of the brain—areas related to EF: attention control, emotional control, working memory, inhibitory control, and organization and planning.[16]

Jazz and EF

A study with middle schoolers demonstrated something important about executive function, but also about jazz. Martin Norgaard conducted a study with 155 seventh and eighth grade students in a suburban Atlanta school.[17] Norgaard and his colleagues conducted a pre-test that measured two EF tasks, one having to do with cognitive flexibility, and one with inhibitory control. The students, who were members of the concert band, were then divided into two groups, each group including both seventh and eighth graders. Both groups had two months of instruction in jazz phrasing, scales, and vocabulary, but only one group, the "experimental" group, was additionally taught to improvise. When they were given tests of EF following the two months of study, the eighth graders in the experimental group had improved significantly in cognitive flexibility, but not inhibitory control. With the seventh graders in the experimental group, it was the reverse. They saw an improvement in inhibitory control, but not cognitive flexibility.

Why the difference? Jazz players all say that in order to be creative, you have to let go of judgments, of preexisting ideas of what and how you should play. The seventh graders had to first learn to inhibit their preexisting ideas about notated music before they could learn to be creative, so they made gains in inhibitory control. Norgaard suggests a couple of reasons why eighth graders improved in cognitive flexibility but not inhibitory control: (1) eighth graders simply had better technique from playing the instrument longer so were better able to engage with the technical demands of improvisation; or (2) as students become more advanced in playing notated music, they tend to be more hesitant at trying improvisation. So for these students, it was more about giving themselves permission to be creative—to be flexible, and cognitive flexibility is where they made gains during the study.

As Bix Beiderbecke, the great American jazz cornetist, pianist, and composer, once said: *One thing I like about jazz, kid, is that I don't know what's going to happen next. Do you?* That's cognitive flexibility.

Music and sound processing

Music can change the world because it can change people.

—singer-songwriter Bono

The study of music has a significant effect on the study of mathematics and on the development of executive functions. But music influences something that is even more fundamental to one's ability to succeed, and that is sound processing. According to Nina Kraus, auditory neuroscientist and the director of the Auditory Neuroscience Laboratory (Brainvolts) at Northwestern University, making sense of sound is one of the most computationally complex tasks we ask our brains to do. Not only is there a staggering amount of information to process, something on the order of 9 million bits of data per second, it must be processed in microseconds to make a response if one is necessary.[18] Making sense of sound isn't all about hearing, as one might assume. A person may have normal hearing and still have trouble processing sound. Auditory processing refers to the listening skills that make it possible for us to make meaning from sound, to understand what is being said in the classroom or meeting room, to understand anger or elation in the tone of someone's voice, to be able to pick out and understand a familiar voice in the middle of a crowded room, say, a teacher's voice in a noisy classroom.[19] A child unable to make sense of what is being heard will not do well in school.

Kraus and her colleagues study the neurobiology underlying speech and music perception, and they have found that our auditory system changes in response to the sounds we hear every day. No one hears the world in quite the same way because each of us has different sound experiences, a different acoustic "footprint."[20] Sometimes the neuroplastic changes in our auditory system are negative, as caused by hearing loss, aging, concussion, or living in poverty (more on that later). But neuroplastic changes can also be extremely positive, as is the case for bilinguals and for people who study and make music. Research done by the Kraus lab on various age groups from toddlers to the elderly has shown that people who study music are better at processing sound, and the refinements that people gain in auditory processing from studying music transfer to other auditory domains, such as speech processing, and then on to language and reading, skills children and adults need to succeed.

In addition to the work that Kraus and her colleagues have conducted in the lab, their research has extended into classrooms and community music programs, real-world studies that demonstrate the benefits of a music education. The lab has also done groundbreaking work with concussion, bilingualism, autism, and aging, but the importance of music-making and music education have figured prominently in Kraus's research, and because of what she has learned, she is a passionate advocate for music education in schools.

Studying music has a positive impact on speech processing

We saw in Chapter 2 that researchers believe our prehistoric ancestors used a proto-musical language with variations in pitch, rhythm, and timbre to communicate and to bond. That proto-musical language eventually evolved into music and speech, and both rely on the crucial acoustical elements of pitch, timing (including rhythm), and timbre to convey information.

Pitch

Any note sung or played on a musical instrument is a combination of the fundamental frequency of that note and a series of other, much softer frequencies called overtones or harmonics, but what one hears and perceives as pitch is the fundamental frequency. The higher the pitch or frequency, the shorter the wavelength of the sound waves. Humans can hear pitches in the frequency range of roughly 20 Hz (Hertz) to 20,000 Hz, although the top range decreases with age. Musicians are all familiar with the concept of A440, the A above middle C on the piano that corresponds to the frequency of 440 Hz. This is the pitch or frequency that orchestras tune to and is known as concert pitch. The range of frequencies produced by musical instruments is roughly 20 Hz to 5,000 Hz. The range of frequencies used by musical instruments is given the letter names of A to G. The lowest note on the piano is A = 27.5 Hz; the highest note on the piano is C = 4,186 Hz, also the highest note of the piccolo. These are the fundamental frequencies. Harmonics, which are part of the sound, extend that range further, up to about 10,000Hz. Speaking-voice frequencies range from about 100 Hz for an adult male to 400 Hz for a small child, so making music involves listening to a far wider range of pitch or frequencies than does listening to speech.

Musicians learn to be increasingly sensitive to pitch as they practice. String, wind, and brass players focus on slight gradations of pitch to be sure they are "in tune" not only with themselves but with colleagues. Singers pay close attention to intonation (or accuracy) of pitch. Pianists play an instrument with a fixed pitch, but they need to be able to tell when the instrument is out of tune, and if they are playing with other instruments or singers, they learn to hear when that person is not in tune with the piano.

In speech, pitch indicates whether we are hearing a statement or a question, it helps us distinguish one speaker from another, and it gives us information about the emotional content of what is being said. In tonal languages such as Mandarin, variations in pitch can totally change the meaning of a given word. Since the frequency of pitches in music is much greater than that used in speaking, and since musicians must pay such close attention to pitch, individuals who have studied music have a greater sensitivity to variations in pitch, and several studies have shown that musicians are better at second-language learning than non-musicians.[21]

Timing and Rhythm

Timing refers to the onset and offset of the sound, the point at which one can perceive the sound and when the sound ends. But when exactly does sound begin? Singers are the only musicians who regularly use the term "onset." For singers, onset refers to the moment that the vibration of the vocal folds begins, and they use different onsets to create different kinds of sound. An onset can be aspirate (soft or breathy), glottal (hard), or easy (gentle, balanced). Saying "he" is an example of an aspirate onset. Holding your breath and then releasing fast while you say "ah" shows you a glottal onset. But if you say a gentle "ah" without first holding your breath, that's an example of a balanced or easy onset.

Instrumentalists often refer to onset as when the instrument "speaks." Different instruments have different types of onsets, depending on how the instrument is made and how the sound is produced. Onsets are often called soft or hard. The piano has a hard onset, and one has no control of the onset. When a piano key is pressed, a hammer strikes the strings. On wind and brass instruments, a player has more latitude in how air is initiated to produce the sound, similar to singers. A wind or brass player can use air and his embouchure (use of lips, tongue, facial muscles, and teeth) to create a hard attack or onset, or a soft attack or onset. On string instruments, the onset will be different depending on whether the string is plucked or bowed and on which of the many kinds of bow strokes are used, such as *legato*, a smooth, connected bow stroke, or *spiccato*, detached notes played with a bouncing bow.

Playing with other musicians involves becoming acutely aware of different onset times. In an ensemble with a piano, violin, cello, clarinet, and soprano, for example, no one consciously thinks about each instrument and voice having a different onset time—after all, it is probably a difference in microseconds. But musicians learn with practice and experience that the sound of each voice or instrument isn't produced at exactly the same time, and they pay attention to subtle cues including (and especially) breathing for information about how to be "in sync." Singers report that nothing is more detrimental to breathing than a pianist who lags just a bit behind. Musicians unconsciously make minuscule adjustments, and as they do so, their auditory systems become more finely tuned.

In speech, each individual sound, syllable, word, and phrase has different timing characteristics or offsets and onsets. A child must be able to rapidly decipher the timing characteristics of each in order to understand what is being said. Hard consonant sounds, such as "ba," "da," or "ga," each have a different onset and are easily confused without accurate speech processing.[22] Timing is also important for hearing when one syllable ends (offset) and another begins. Timing, in addition to onset and offset, also refers to the ability to synchronize with the beat or with a particular rhythm. Rhythm is the placement of sounds in time and refers to a pattern of strong and weak beats. Being able to keep time to a beat is a fundamental skill for music performance, but it is also fundamental for speech processing.

Timbre

It is sometimes said that timbre is what is left of the sound that isn't pitch or loudness. Also called tone color or tone quality, timbre refers to the characteristics of sound that make one voice sound different from another or one instrument sound different from another. One can tell the difference between a clarinet and a violin playing the same pitch because each has a different timbre. Timbre is determined primarily by the harmonic content of a sound. All pitches have the same harmonic series, but the different design of each instrument means that the frequencies in the harmonic series are balanced in a different way, so the sound color or timbre of the instrument varies. A singer's instrument is the body, and the body parts involved in singing—the lungs, windpipe, larynx, tongue, lips, and mouth cavities—are all unique to that person and create a particular timbre. The harmonics at the beginning of a note, the attack or onset, are particularly important for timbre. It is easier to identify differences in timbre between two instruments or voices if the articulation of the sound is clearly heard than if only a long, smooth sound of the pitch is heard without the initial attack.

Musicians learn to distinguish the *timbre* of one instrument or voice from another in an ensemble, even if they are playing or singing the same pitch. And each instrumentalist or singer learns techniques to change the timbre of the sounds he is producing. For example, string players learn the difference between the timbre of the pitch of an open string or fingered string, and the difference in timbre when the pressure or speed of the bow is changed. Brass and string players produce different timbres by muting the instrument. A pianist can use the una corda, damper, and sostenuto pedals to change timbre. Singers learn to change timbre to make the voice lighter or darker, breathier, more mellow, or warmer by changing the shape of the vocal tract.

In speech, timbre gives us information about the speaker. Just as with musical instruments, two voices may be speaking at relatively the same pitch, but the timbre tells you immediately if the voice belongs to your best friend or to your cousin. An individual speaking voice can also change timbre slightly, as does the singing voice. You may remember from Chapter 3 that mothers throughout the world tend to shift the timbre of their voice when they speak to their babies. One mother may shift to a breathier tone, another may be slightly more nasal, but it is a specific timbre for her baby, and the babies recognize the shift from a normal speaking voice to their personalized timbre.

Timbre also gives us info about *phonemes*, the individual sounds that make up words, such as p, b, d, t, k, etc.[23] As one speaks, the shape of the vocal tract changes slightly as the articulators (lips, teeth, tongue, jaw, palate) form different phonemes. As with instruments or singers, that changes the balance of the harmonics, and the timbre changes slightly between letter sounds or phonemes. Children need to be able to distinguish the minute differences between these sounds, such as the "b" or "p" in "bat" or "pat," or the "d" or "t" in "bad" or "bat," to support early literacy and language development.

Timing and timbre have to do with the verbal message itself, with what a speaker is saying. Pitch has to do with the intent of the message and the emotional content.[24] Although sensitivity to pitch, timing, and timbre constantly improve the longer one studies music, even a beginner, especially in a group setting, learns to be sensitive to these elements. He must learn to play in time with the group, follow a conductor, learn to tune to other instruments, learn the difference between the sounds of a clarinet and a saxophone. Even a beginner learns how to make a simple tune sound happy or sad. Sensitivity to pitch, timing, and timbre begins early in a music education. Brain imaging showed earlier or larger responses to musical tones in twelve-month-olds who had just had six months of interactive music lessons with their parents.

Their brains were already showing neuroplastic changes in response to inter-active music play with parents (see Chapter 3).

We saw in Chapter 4 that when cognitive, sensorimotor, and reward sys-tems in the brain are all engaged at the same time, neuroplasticity occurs more readily, and all three are engaged when making music.[25] *Sensorimotor networks* in the brain are activated when sounds connect with the motor actions necessary to make those sounds. *Cognitive networks* are activated when attention is focused on the elements of pitch, timing, and timbre, and when working memory is used in sight-reading or incorporating new ideas into what has already been learned. Cognitive networks are also used when sound is connected to meaning—when one learns, for example, that a dominant-seventh chord has a certain unstable sound and wants to resolve to a more stable chord, or a minor key sounds sad or melancholy compared to a major key. Making music causes feelings of happiness or pleasure, activating the *reward system* in the brain. Whenever the reward system is activated, we want to repeat that activity, thus driving neuroplasticity.

Making music involves all three systems and has been found to be a par-ticularly strong driver of neuroplasticity. Because music and speech share the same auditory pathways, and because music requires finer auditory distinctions for pitch, timing, and timbre, the enhancements gained in the auditory system through music-making transfer to the neural processing abilities necessary for speech, then to listening and language skills.[26]

Psychologist Aniruddh Patel has proposed a slightly different framework to explain why musical training can enhance speech processing.[27] Called the "OPERA" hypothesis (easy for musicians to remember), his theory suggests that studying music can benefit speech processing when all five of the fol-lowing conditions are met: music and speech processing *overlap* in the brain circuits; music places higher demands or *precision* on these shared networks; music elicits *emotion*, which drives neuroplasticity; practice includes much *repetition*, which also drives neuroplasticity; and practice requires close *attention*.

How auditory processing is measured in the brain

Until recently, the usual description of the auditory pathway has been as a one-way path from ear to brain, as it was described in Chapter 4. But according to Kraus's research, auditory processing isn't a one-way path, it is an interactive ear-to-brain and brain-to-ear system.[28] At every point along the path from ear to brain, the auditory signal is met by signals coming from the cerebral cortex, signals having to do with attention, working memory, and sound-to-meaning connections—the brain-to-ear system—and that shapes what we ultimately hear. Over time, our response to any incoming sound becomes fundamentally altered due to previous experiences with sound, our memories for familiar or previously heard sounds, and the meanings we assigned to certain sounds—in other words, how we hear is influenced by our lives in sound.

The thin gray arrows in Figure 10.1 show the auditory pathway from the inner ear to the brainstem and continuing to the auditory cortex. Signals from cognitive areas of the brain having to do with attention, working memory, and sound-to-meaning connections that interact with the incoming signal are shown in thicker black lines. Kraus uses an analogy of a mixing board and its ability to fine-tune certain auditory signals to describe what happens in the auditory brainstem.[29] In a sound studio, an audio engineer mixes audio signals from two or more sources (e.g., multiple microphones recording different instruments or voices). He may boost or cut certain frequencies to improve the sound, minimize unwanted sounds, and regulate volume levels and sound quality to produce a good output signal that is broadcast, recorded, or amplified through a sound system.

At the brainstem, the incoming signal from the ear may be good quality, or it may be degraded due to hearing loss, autism, noisy environments, or concussion, just as an incoming signal to a mixing board may vary in quality. The signals from cognitive areas of the brain basically tell the brainstem what to pay attention to, augmenting some sounds, excluding irrelevant information, controlling for context—what has just been heard that has a bearing on what is now being heard (working memory). Those signals, which are like the audio engineer at a mixing console, will be different in each individual based on experience, and they also vary in quality. Children who grow up in poverty, for example, have a severely compromised incoming sound signal as well as inefficient signals from cognitive areas of the brain (more on that later).

Musicians, on the other hand, have very finely tuned brain-to-ear signals. They are better at picking out the most meaningful parts of the sound. People who have studied music are accustomed to separating out one voice

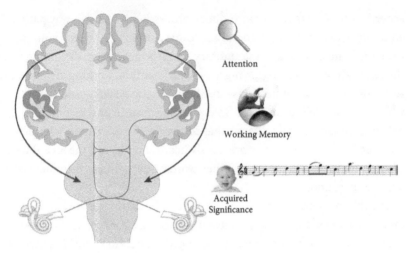

Adapted from Kraus and Chandrasekaran (2010) *Nature Reviews Neuroscience*

Figure 10.1 The human auditory system
Credit: Reprinted by permission. D. Strait and N. Kraus, "Playing Music for a Smarter Ear: cognitive, perceptual and neurobiological evidence," *Music Perception* 29, no. 2 (2011), University of California Press.

from many, to paying attention to gradations of pitch, timing, and timbre, to keeping notes or rhythms in working memory, and to the subtleties of expressing emotion, so those circuits are enhanced in musicians and thus have a positive effect on the incoming speech signal. The musician's brain is better at picking out the relevant portions of the sound, as a skilled engineer can do at a mixing board. And they will be better at making sound-to-meaning connections because everything they do when studying music is about attaching meaning or significance to the sounds they make and hear. For example, they learn what a major key sounds like as opposed to a minor key, what makes consonance and dissonance sound different, that stretching a few beats (rubato) can add emotion to a musical excerpt, or that playing faster adds excitement. Meaning, or significance, is attached to musical sound. The infant depicted in Figure 10.1 has already learned to connect the sound of a certain kind of melody with happiness.

Musicians themselves will have different strengths. A conductor has better spatial auditory processing than a pianist because he is accustomed to separating out streams of sound across the expanse of the orchestra.[30] Musicians who are improvisors or aural learners have faster neural processing than non-improvisors.[31]

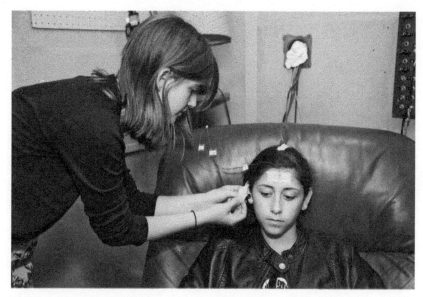

Figure 10.2 Brain responses are captured by sensors placed on the head; sound is delivered via earbuds

Credit: Auditory Neuroscience Lab, Northwestern University

Kraus's lab devised a way to measure neural responses at the brainstem by attaching electrodes to the forehead and scalp of the subject to record neural responses to speech and music, as seen in Figure 10.2. Although the response is measured at the brainstem, it is influenced by the vast interactive ear-to-brain, brain-to-ear circuitry. The recorded response is an objective marker of auditory function, and it gives the researchers information about how details of sound are transcribed in the brain and how an individual's brain is processing sound. These electrical responses can be read by a computer as a waveform that provides information about pitch (frequency), timbre (harmonics), and timing, as seen in Figure 10.3. The waveform also gives information about the consistency of the response and about the amount of neural noise (brain neurons firing spontaneously even when there is no sound). The less neural noise, the better the sound quality. Amazingly, if the electrical response recorded at the brainstem is played as a sound file, it sounds eerily like the original sound source *[See the companion website at item 10.4 for a link to a video "Our Biological Approach," by the Brainvolts Lab. The brainstem response played as a sound file is within this video ⦿.]*

Figure 10.3 Electrical activity in the brain in response to sound can be visualized
Credit: Auditory Neuroscience Lab, Northwestern University

Positive effects of studying music on sound processing

The Kraus Lab has done a great deal of research concerning the positive effect of studying music on sound processing. Following are a few highlights:

Brainstem responses to pitch, timing, timbre
Musicians have stronger brainstem responses to both speech and music than do non-musicians, and the strength of the response correlates with the number of years of practice. Musicians show super-accurate pitch encoding as well as enhanced transcription of both timbre and timing cues necessary for both speech and music.[32]

Connection between rhythm skills and reading
Preschoolers between the ages of three and four who can sync to a beat have more accurate brainstem processing of speech than those children who cannot synchronize. They also test higher on tests of early language skills, such as auditory short-term memory and rapid naming of visual symbols, skills related to language and reading.[33] The results of this study

with preschoolers suggest that we should be playing rhythm games with preschoolers, helping them with beat synchronization to prepare them to be better readers, as well as better musicians.

Preschoolers aren't the only age group to develop better reading skills through beat synchronization; teenagers do also. Teenagers who are better able to sync to a rhythmic beat have more consistent brain responses to speech. This means that they will have stronger sound-to-meaning associations that are crucial to reading ability. To distinguish speech sounds, a listener must be able to detect extremely small differences in timing—onset times. Practicing syncing to a beat leads to more stable representation of sound in the brain, leading to better linguistic and reading skills.[34] Interactive drumming, drum ensembles, or drum circles are all enjoyable opportunities for teenagers to improve their brain responses to speech and thus improve reading ability. (We will see in the Epilogue that syncing together through music also leads to more pro-social behavior.)

Hearing speech in noise

Although this may come as a surprise to many musicians who are bothered by noise in a crowded restaurant, researchers in the Kraus lab have found that musicians process speech sounds in a noisy environment better than non-musicians.[35] This is important for children in noisy classrooms, or for adults in noisy meetings or restaurants. Musicians, even at an elementary level, learn to pick out one sound out of many, to hear the melody embedded in the harmony, to hear how their own part fits into the ensemble. And because of that experience, musicians have more finely tuned signals from the cortex telling the auditory signal at the brainstem what part of the sound to pay attention to and what to filter out. Working memory tells the signal at the brainstem what we just heard that provides context. Musicians have stronger encoding of speech harmonics, the timbral part of speech, and that helps to identify specific voices in a noisy environment and separate one voice from another. Studying music can strengthen the very same processes that are impaired in children with dyslexia or other learning disorders who often have difficulty hearing in a noisy environment.[36]

Hearing emotion in sound

Musicians have enhanced processing of emotion in sound, and the ability is related to the number of years of experience of the musician and the age at which study began. Musicians are not only more sensitive to the complex part of the sound that has the most to do with the emotional content (such as a baby's cry), they also de-emphasize the simpler or less emotional component of the sound.[37] Musicians are not aware of doing this, but since they practice using subtle changes in timing, volume, or timbre to convey emotion in a piece of music, that sensitivity carries over into auditory processing ability in other situations as well. Being able to identify emotion in speech is a skill that is useful in our personal relationships, in classrooms, business settings, and everyday encounters with people.

Dana Strait, primary author of the study, suggests that musical training might promote better emotion processing in individuals with autism or Asperger's, and in fact, later studies show this to be the case.[38] The Kraus Lab has found that the aspects of sound that tend to be diminished in individuals with hearing disorders, autism, dyslexia, concussion, or living in poverty, are the same parts of the sound that are enhanced by music training.

Effect of studying music on aging

Studying music early in life has benefits as we age. Musicians aged forty-five to sixty-five who had played an instrument since they were nine or earlier were better than non-musicians at hearing speech in noisy environments and had better auditory working memory. This means that studying music may reduce the impact of auditory decline as one ages.[39] And even though they had not studied for forty years, adults aged fifty-five to seventy-six who had studied music for four to fourteen years as children had faster responses to speech sounds than did those without training.[40] So even a little music training early in life has lifelong advantages.

Sound processing and living in poverty

Researchers have known for some time that living in poverty has a negative effect on a child's brain development, learning, and academic performance. Children in poverty often have inadequate nutrition, insufficient language stimulation, and little to no attention or emotional support. A study from the University of California at Berkeley shows that parents experiencing financial insecurity talk less to their children, so those children just don't grow up hearing very much vocabulary. The study concludes: "if you are worried about putting food on the table tonight, or scraping together money for that medical bill, or figuring out where to enroll your child in school now that you have been evicted from your neighborhood, you may be less likely to narrate the color of the sky to your child as you ride together on the bus."[41]

Poverty is often compounded with neglect because one or both parents are working more than one job and don't have the time or emotional resources to deal with their children. Impoverished, neglected children often show cognitive deficits that include impaired executive function, attention, processing speed, language, memory, and social skills.[42] In addition, these children often live in areas where there is environmental or noise pollution, and prenatal environmental pollution has been found to have a negative academic effect on children by the time they are adolescents, with these children demonstrating poor skills in spelling, reading comprehension, and math.[43]

Kraus and her team have found that children who grow up in low socioeconomic areas (i.e., in poverty) have more "neural noise" in their brains than their classmates, even when no external sound is present. Neural noise refers to brain neurons firing spontaneously in the absence of sound—and this slows auditory growth. Neural response is erratic, with lower fidelity to the incoming sound. Kraus says neural noise is like static on a radio—with the announcer's voice coming in faintly. These children's performance on tests of reading and working memory is poorer than that of their peers who do not live in poverty.[44]

Poverty is far too prevalent in America. The Annie E. Casey Foundation reported in December 2020 that one in seven families with children (14%) reported that they had not had enough food in the previous week, and one in five households with children (18%) didn't think they would be able to meet their next mortgage payment. Those numbers are significantly higher for Black, Latino, and mixed-race households. These children are at a significant disadvantage before they even begin school.[45]

Kraus wondered if opportunities for music education might offset the biological impact of poverty. Two research studies conducted in real-world situations, one in the Chicago Public Schools, and one concerning Harmony Project, a community music program in gang-reduction zones in Los Angeles, provided the answer. She calls these studies her neuroeducation work and considers them to be cornerstones of her research. Kraus says that she wanted to study the effects of musical experience on the nervous system in a real music program, not one that was created by scientists. She wanted to know "what is the effect of something that real musicians do in a real school, in music programs that have a history of working for a decade or so."[46]

Neuroeducation

Chicago Public Schools study

Kraus and her colleagues designed a study involving at-risk kids from low-income neighborhoods in Chicago. They wanted to see what would happen if music training were begun as late as high school. Would it still have a beneficial effect on sound processing? They followed two groups of high school students from Chicago schools in low-income neighborhoods for three years. One group was enrolled in band classes that met for two to three hours a week of instrumental instruction, including instruction on technique, sight-reading, performing in an ensemble, and regular assessments of progress. The other group was enrolled in Junior Officer Reserve Training Corps (JROTC), which involved classroom instruction, fitness-based training, and included a performance component. Both groups met for the same amount of time each week and were educated in the same classrooms.

Participants were tested before entering high school for neural processing and language abilities. They were tested again three years later just before entering their senior year of high school. The results showed that even if music training is begun as late as high school, it has a positive effect on the brain's response to sound and can therefore have benefits for language skills. The same was not true for the JROTC students.[47] Kraus points out that, although both groups were involved in what are known as enrichment activities, it wasn't just any enrichment activities that increased brain development—it was specifically music study.

Harmony Project

There are an increasing number of nonprofit music organizations in the United States that provide music lessons, instruments, ensemble experience, and performance opportunities for children from low-income families and underserved neighborhoods. The organizations have missions that range from empowerment of students to community development, increasing access to higher education, and fulfilling a need for students who desperately want to make music but do not have the financial means to do so. Consistently, these organizations report that students who spend several years studying music tend to do well academically and most go on to college *[See item 10.5 on the companion website for links to Harmony Project, discussed here, as well as to several other successful community music organizations ⊙.]*

One of the most successful community music programs is Harmony Project, founded in 2001 in Los Angeles by Dr. Margaret Martin. It's also the first program to work with neuroscientists to determine what is happening in the brains of these disadvantaged students that leads them to excel academically as they study music. Harmony Project demonstrates the impact one person can have on the lives of thousands of children.

One day in the late 1990s, Martin was at the Hollywood Farmers' Market with her five-and-a-half-year-old son Max. Max was a violin prodigy who had taken out his violin and was playing Brahms, and as happens with street musicians, listeners were dropping money in his violin case. Precocious Max, according to his mother, liked to make money.[48] Martin nervously watched as a group of tattooed gang members walked up to him and stopped in front of his open violin case full of money. Were they going to take money from a five-and-a-half-year-old? They listened attentively for a while, and then pulled money out of their own pockets to carefully put in his violin case. She wished at that moment that she could have given those gang members other opportunities to create better lives for themselves. Perhaps they might also be playing the violin. Eighteen months later, Martin, whose graduate degrees are in public health, founded Harmony Project, a nonprofit organization that provides instruments, lessons, orchestras for the students, field trips, concerts, and college scholarships for children from underserved neighborhoods.

Harmony Project quickly grew in popularity, not just because students thrived musically, but because they thrived academically as well, and the waiting list to get into the program grew. Martin noticed that Harmony Project students were graduating at the top of their classes, and most were going on to college where they excelled. She suspected that something was going on in their brains that primed them for academic excellence but didn't know how to prove it. She cold-called the head of Child Health and Human Development at the National Institutes of Health, who just happened to have recently heard Nina Kraus speak at a conference and knew that Kraus's research interests dovetailed with what Martin was trying to find out about the brains of the Harmony Project students.

In the summer of 2011, four researchers from the Kraus lab flew to Los Angeles to begin their research study of Harmony Project students. Their subjects were seventy-five second graders who wanted to enroll in Harmony Project. The researchers conducted an extensive series of tests to determine hearing abilities, memory, attention, language, and the neural responses to speech using their method of recording responses at the brainstem. Kraus and Martin had determined that half of the students would immediately begin instrumental lessons. The other half would participate in general music classes for a year, listening to music and learning basic note reading and music history. After a year, the researchers returned to Los Angeles in summer 2012 and repeated the tests. As they had expected, they found no changes in cognition, language, or brain function in the students who had taken the general music classes with no instrumental instruction. But to their surprise, they also found no changes in the children who had been studying an instrument for a year.

Martin was undeterred by the test results and suggested they come back the following year, and they returned in the summer of 2013. At that point, the children who had studied musical instruments for two years showed significant increases in neural sound processing. Their language and listening skills were above those of their peers, and the degree of improvement was related to how engaged they had been in the program, whether they had attended all the lessons and rehearsals. This is not unlike the results of the Bergee and Guhn math studies that showed that it was achievement in music that predicted math success, not just participation. Here, it was the degree of engagement—and achievement—in the activities of Harmony Project that determined improvement in sound processing. As we have seen with Kraus's previous work in the lab, speech processing and language skills improved in children studying music, but this time it was in a real-life, community setting.[49]

Further proof that studying music improves academic outcomes

With the children at Harmony Project and the teenagers in the Chicago Public Schools, Kraus and her colleagues demonstrated that studying music improves neural encoding for speech for at-risk students, and improved neural encoding sharpens their hearing and language skills.

Further proof comes from the students themselves. One hundred percent of the Harmony Project students are in a Title 1 school (federal funds provided to schools with high percentages of low-income students) and/or are enrolled in the National School Lunch Program. The vast majority (over 95%) are African American, Latino, Asian/Pacific Islander, or biracial. These two factors alone make them high-risk *[These statistics and those that follow are from the Harmony Project 2020 Annual Report, which contains other interesting information as well. See item 10.6 on the companion website ⊙.]*

But at Harmony Project, students are given an instrument to use at home, and they spend between four to ten hours a week in lessons and ensembles, provided they remain enrolled in school. Graduating students have spent an average of seven and a half years in Harmony Project. The LA Chief of Police selected the Los Angeles program sites, and they are in gang-reduction zones, high-crime areas.[50] The dropout rates in the high-crime areas where the students live are 50 percent or higher, yet more than 90 percent of high school seniors who have been in Harmony Project for at least three years have gone to college, and there is a 97 percent college acceptance rate. Seventy-seven percent of these students are first-generation college students, and 54 percent of Harmony Project alumni have earned a bachelor's degree by the age of twenty-four, compared to the national average of 12 percent and 13 percent for African American and Latino students. Harmony Project students are being accepted by, and attending, schools such as Harvard, Princeton, Georgetown, Dartmouth, NYU, Stanford, UC Berkeley, UCLA, and others. Two students have received Fulbrights, one has received the prestigious Gates Millennial Scholarship, and many go on to receive graduate degrees. Perhaps not surprisingly, almost half of them go into STEM fields, the same fields that extensively use spatial-temporal reasoning, which, as has been demonstrated earlier, is enhanced in individuals who study music. For a program that is only twenty years old, the level of success is remarkable.

The achievements of Harmony Project have not gone unnoticed. In 2009, founder Margaret Martin received the Coming Up Taller award from First Lady Michelle Obama at the White House. The award recognizes programs targeting children who traditionally lack access to the arts and humanities. Two years later, Martin went back to the White House to receive the Presidential Citizens Medal from President Barack Obama. This is the second highest civilian honor the US government bestows. In 2015, the US Department of Education designated Harmony Project a "2015 Bright Spot in Hispanic Education."

While awards may be external validation, it is the success of the students that makes Martin proud of, and passionate about, the program she founded. There are currently sixteen sites in Los Angeles County and southern California and eight affiliates in other states, and Martin is committed to expanding the program to every low-income area in the country. Every child is born with musical abilities, and all children deserve access to a music education.

The power of music to increase human potential

The Harmony Project students demonstrate that speech processing and language are enhanced by studying music. Since so many of them enter STEM fields, it would also suggest that the spatial-temporal abilities they have gained through studying music transfer to math, and executive function skills that are improved through the study of music lead to academic success in many areas. And in fact, a recent study by researchers from Los Angeles and the University of Vermont looked at academic improvement of 1,080 second grade children enrolled in Harmony Project in five elementary schools in the Long Beach Unified School District in California (LBUSD). Harmony Project at LBUSD was founded in 2014, and like the original Los Angeles program, is located in high-poverty elementary schools in high crime areas. Over a two-year period, students who had enrolled in Harmony Project and participated in four hours of instruction per week had higher Achievement Rating scores in math, reading, writing, and speaking than their classmates who were not enrolled in Harmony project. The greatest improvement in the reading and math scores was seen in students who had the lowest achievement scores prior to enrolling in the program. The authors of the study point specifically to the achievement in speaking ability as being "critical to student agency, resilience, and success."[51] Students from Harmony Project LBUSD can be seen in Figure 10.4.

The impact that studying music has been shown to have on literacy skills is substantial. Yet opportunities to study music within the public school curriculum are rapidly diminishing, especially in low-income communities. According to the National Endowment for the Arts, between 1982 and 2008 the percentage of African American young adults who received arts education in childhood dropped from 51 percent to 26 percent. For Hispanics, it was 47 percent to 28 percent, for Whites, 59 percent to 58 percent. By 2010, the most recent year statistics were available, only 23 percent of high school seniors participated in music or the performing arts.[52] Budget cuts across the country have made that number drop every year as more schools cut music programs.

Figure 10.4 Harmony Project students from Long Beach elementary schools
Credit: LBUSD Harmony Project

The question shouldn't be "Does music make you smarter?" but "How can we ensure that every child has access to studying music?" The Sounds Academy in Phoenix, a nonprofit music education organization, points out: "The zip code of a child should not dictate their access to a music education." Our prehistoric ancestors were making music tens of thousands of years ago and our brains have evolved to support—and to value—music-making. Music is a basic part of our humanity, and every child should have the opportunity to study music, not because it may give him cognitive advantages in another academic area, but because the ability to engage with music, as with language, makes us more fully human.

Key Concepts

- There is no evidence that listening to music makes one smarter, but there is considerable evidence that learning to play a musical instrument and making music gives lifelong cognitive advantages in many areas.
- The idea that training or developing neural resources in one area or domain influences other domains is called transfer. Studying music has both a near transfer (closely related to music) and a far transfer (non-music related) effect on other areas of study.
- Studying music has a far transfer effect on the study of mathematics. Music and mathematics have shared neural resources having to do with

the meaning of symbols and the mental manipulation of those symbols, in other words, spatial-temporal thinking.

- The study of music has a far transfer effect on executive function skills: inhibitory control, working memory, and cognitive flexibility. Good executive function skills are important in classrooms, jobs, and personal lives.
- The human auditory system changes in response to the sounds heard in daily life, and studying music has a far transfer effect on sound processing abilities.
- Musicians have heightened responses to the acoustic details of speech (pitch, timing, timbre). Because the same auditory pathway is used for both music and speech, the enhancements from studying music transfer to speech processing and language.
- Musicians have faster responses to sound, are better at hearing speech in noise, and are more sensitive to the emotional content of speech.
- Community or in-school music classes can improve the neural encoding of speech for at-risk children, leading to better language development and reading, and thus to academic success.

Epilogue

Thoughts on Music and Society

The Philadelphia Orchestra had planned to begin a major celebration in March 2020 in honor of Beethoven's 250th birthday. The first concert was scheduled for March 12, all nine symphonies to be performed over a four-week period with the cycle repeated later at Carnegie Hall. But on March 11, the World Health Organization declared COVID-19 a global pandemic and performing arts organizations throughout the world closed their doors. In response, the following evening the orchestra performed Beethoven's Fifth and Sixth Symphonies to an empty hall, reaching more than half a million people through livestreaming and airing on public radio. It was an emotional experience for orchestra and listeners alike because no one knew what would happen in the next few weeks or when the orchestra might be able to return to performing for live audiences.

A week later, Music Director Yannick Nézet-Séguin spoke with Terry Gross on *Fresh Air*, her National Public Radio program. Gross asked Nézet-Séguin how he felt about reports that people listening in their homes had been texting each other during the concert. Nézet-Séguin's answer was perhaps surprising to many. He didn't lament the fact that people were texting, instead he commented that performing music live is about building community and sharing the music. That usually happens in the concert hall, but if that's impossible, then the need to share in the event can be accomplished by texting with others listening to the same performance. He supported the fact that audience members sitting in isolation at home would reach out to friends who were also listening and share the emotional experience of hearing these great works by Beethoven.[1] *[A link to this interview is found on the companion website, item E.1 ⊙.]*

Learning to sing or play an instrument is an individual experience. This book has considered the brain processes that not only make that possible

The Musical Brain. Lois Svard, Oxford University Press. © Oxford University Press 2023.
DOI: 10.1093/oso/9780197584170.003.0011

but also make it possible for individuals to become superb performers. However, music is also a shared social glue, and perhaps nothing has demonstrated this so vividly on such a wide scale as the coronavirus pandemic. The need to share music and feel connected to others was evident everywhere. In the early weeks of the pandemic, the Italians sang together from separate balconies. Five BBC radio stations simultaneously broadcast the same program, so people throughout the UK could sing along and know that they were sharing the experience with others. The Acapella app, although available for several years, suddenly became widely popular with small groups who used it to record parts separately and appear together on screen in a grid. Cellist Yo-Yo Ma created a community through #songsofcomfort, posting a video of music that gave him comfort and asking others to do the same. In response, people of all ages and abilities uploaded and shared performances. As the pandemic dragged on and 2020–2021 concert seasons were canceled, performing arts organizations found creative ways to make music in small groups, share those experiences, and thus build community. There was an explosion of online musical content. Making and/or experiencing music together, no matter the level of expertise, was important to combat pandemic isolation, helping people to feel comforted and connected.

Why is music a social glue?

Why do we need music to feel connected? Our prehistoric ancestors needed their proto-musical language to survive, to build trust and cohesion, to cooperate in the tasks necessary for daily living. But they also made flutes that were undoubtedly played as they gathered to celebrate a hunt or for ceremonial or religious purposes. Our brains have evolved not only to support but to value music-making as a group. Playing music or singing together leads to people feeling closer to one another, to increasing sense of community, to being more cooperative and helpful to others, and to strengthening social bonds.

The brain is an electrochemical organ, the billions of neurons communicating via electrical signals and chemicals. Researchers have looked at neurochemistry in the brain and at electrical activity in the form of brain waves to explore the link between music and social bonding. They have also looked at the implications for social behavior when people physically sync their movements, which happens when making or listening to music together.

Neurochemistry

Oxytocin is a hormone usually associated with childbirth and breastfeeding but also with empathy, trust, relationship-building—and singing. Researchers have known for some time that group singing promotes trust and cooperation among members of the group.[2] They now know the increased trust and cooperation is due to oxytocin being released during singing. Oxytocin levels have been shown to increase in both professional and amateur singers after a thirty-minute singing lesson, and group singing has also been found to increase levels of oxytocin in group members.[3] As oxytocin is released in individuals who sing together, they become more trusting and cooperate more with others in the group, whether or not they have anything in common. If those same individuals are chatting, there is no increase in oxytocin levels and therefore no impact on relationships.[4]

Dopamine is the "feel good" neurotransmitter and is released with biological rewards like food and sex, and with addictive drugs like cocaine, making us want to seek these things out again. But dopamine, part of the brain's reward system, is also released when making or listening to music we like as well as during social interactions. So when we listen to or make music with others, dopamine is released both in response to the music itself and to the social aspect of listening with others, triggering the brain's reward system and making us want to share the experience of listening to music with others again.[5]

Cortisol is known as the body's stress hormone, and levels of cortisol increase to boost our energy to help us deal with the stress. For our prehistoric ancestors, that may have meant increased energy to escape wild animals. Today, extra cortisol is likely to be released when we experience a work- or family-related crisis or when we experience the stress of performing—performance anxiety, with symptoms of shaky fingers, increased heart rate, the "fight or flight" response. On the other hand, cortisol levels decrease when one is listening to music. Volunteers who listened to a live concert of choral music experienced significant reductions in cortisol levels.[6] One becomes more relaxed, which is why so many people wanted to listen to music during the pandemic.

Electrical signals and brain waves

When billions of neurons communicate with each other via electrical impulses, those impulses synchronize; the synchronized electrical activity is called a brain wave. Brain waves are detected by EEG (electroencephalogram) by placing electrodes on certain places on the scalp to record electrical impulses in the brain. They are measured in cycles per second, or hertz (Hz), the same designation used for pitch (see Chapter 10). And just as different frequencies, or Hz, are associated with different pitches, they are also associated with different brain waves. Gamma waves are the fastest, 39–42 Hz, and indicative of peak concentration. Beta waves (13–38 Hz) are associated with mental alertness, having an active conversation, or being engaged in your work. Alpha waves are slower (8–12 Hz) and are associated with relaxation. Theta waves (4–7 Hz) are associated with daydreaming, and delta waves (1–3 Hz) are associated with sleep. We saw in Chapter 6 that consolidation of declarative memory is associated with slow wave sleep, characterized by delta waves.

Making music together affects the repetitive patterns of neural activity in the brain, seen as brain waves. When two musicians play together, their brain waves synchronize, even if they have not played together previously. While one might think this is logical because musicians try to be in sync when they play together, the syncing of the brain waves begins even before they begin to play. Surprisingly, the brain waves that sync are the low-frequency theta and delta waves, slow waves associated with deep meditation and sleep, below the level of consciousness. These brain waves have also been found to be involved in social interactions, from mother-infant bonding to "theory of mind," or the ability to understand the behavior of others.[7]

It's not just the brain waves of performers that sync together. Brain waves of listeners synchronize to each other when they listen to music as part of a group. Listeners' brain waves sync more when they listen to live performers than to a recorded video of the same concert, and the more in sync listeners' brain waves are, the more connected they feel to the performer and the more they enjoy the performance.[8]

Brain waves also sync between performer and audience—in areas of the left hemisphere having to do with empathy and in areas of the right hemisphere having to do with recognizing musical structure and pattern. In the right hemisphere, they are also found in areas involved in interpersonal understanding. Perhaps not surprising, these are also the areas where mirror neurons are found, the brain cells that help us understand the actions and emotions of others (see Chapter 9). The more the audience members reported enjoying a particular work, the more the brain activity was found to be synchronized with that of the performer.[9]

Synchronization of movement

Both neurochemicals and brain waves have been implicated in the connection between music and social bonding. That connection has also been shown in synchronization studies. An area within the basal ganglia called the caudate has been found to be involved in the processing of synchronization, our ability to synchronize both with another person or with an external beat. The caudate also plays an important role in facilitating pro-social behavior. Perhaps not surprising, then, participants who synchronized with a partner while drumming were later more helpful in a helping task, and the amount of activity in the caudate correlated with how helpful they were. The effects were stronger in the participants who were able to synchronize more easily.[10]

Students who sang in unison reported stronger feelings of being on the same team than students who were singing but not in sync, and when they played an economics game that they didn't know was related to the study, synchronizing singers made greater contributions to an account benefiting the group at a direct cost to themselves, so singing in sync directly led to behavior benefiting the group rather than the individual.[11]

Social behavior related to syncing begins at a very early age. The passive versus active listening study described in Chapter 3 demonstrated that six-month-olds participating in interactive music classes with their parents showed better early communication skills than did infants who had only listened to music. At that age, they were unable to sync, but there was something about trying to make music together that made these infants happier and more communicative.

In another study, a group of four-year-olds were randomly assigned to either a music group or a no-music group. The music group interacted with each other and an adult in the context of singing, dancing, and playing musical instruments to an easy-to-learn children's song. The no-music group interacted with each other and an adult while listening to a story. After these activities, all the children played two games, a "Cooperation" game and a

"Helping" game. The children who had been making music together were significantly more cooperative and helpful.[12]

Moving together to music seems to encourage everyone to be more helpful—even toddlers. In a study at McMaster University, fourteen-month-old infants who were bounced in synchronization to "Twist and Shout" with a researcher facing them were more likely later to help that researcher.[13] One researcher held a baby in a forward-facing baby carrier and stood facing a second researcher. When the music began, both researchers began to bounce up and down, one holding the baby. Some of the babies were bounced in sync with the music and others were bounced out of sync.

When the music stopped, the researcher who had been facing the baby began a task in which she pinned dishcloths on a clothesline and "accidentally" dropped a clothespin. The babies who had been bounced in time with the researcher facing them were far more likely to toddle over to the dropped object, pick it up, and hand it to the researcher. The babies who had been bounced out of sync weren't interested in helping and walked away. This wasn't a single instance; there were thirty toddlers in the study. Being helpful after bouncing in sync to music was the norm. *[A video excerpt from this study provided by the Auditory Development Lab at McMaster University can be seen as item E.2 ⊙.]*

Multiple research studies involving neurochemistry and electrical activity have shown what happens in the brain when we make music together; behavioral studies show how behavior is affected after syncing with others to music. But as has been the case throughout this book, the stories about real people are perhaps the best demonstration of, or confirmation of, the research.

The importance of music for creating community

Stranded in the Antarctic

The story of Ernest Shackleton is quite well known. He was the leader of the 1914 expedition to cross the Antarctic by way of the South Pole. His ship, the *Endurance*, became trapped in an ice pack, and the men eventually had to abandon it and strike out across the ice on foot, dragging their lifeboats behind them with ropes. Shackleton ordered that only the bare necessities could be taken; they each had a two-pound personal allowance. They left behind money and jewels, clothing, books, and even scientific instruments because they would no longer be able to gather data. They took along medical supplies, photographs, diaries—and Leonard Hussey's five-string banjo. Hussey was the meteorologist on the crew, and he had played his banjo frequently on board. The banjo weighed twelve pounds, well over the two-pound limit, but Shackleton ordered that the banjo be taken along. He called it "vital mental medicine," and knew it would be important to help with the depression the crew would face. And he was right. Many of the men kept diaries and wrote, often with humor, about the positive effect of the banjo on their mental state.[14]

It's unclear how or why Shackleton knew the banjo and music-making would be of such great importance to his men, but his decision was wise. The men were in daily danger not only of being unable to find food, but also of losing their lives. There was a tremendous sense of isolation as they looked in every direction and saw nothing but ice. The music they sang together with the banjo created a sense of community that helped them complete their 346-mile hike across the ice. All the men—and the banjo—survived.

Making music together built trust and cooperation among our prehistoric ancestors, and it facilitated trust and the building of community among Shackleton's crew, enabling them to survive. But sometimes survival isn't strictly physical, it's emotional as well.

The Dallas Street Choir

The homeless are a particularly vulnerable population. Homeless individuals feel a profound sense of isolation and hopelessness; survival is a daily concern. Many consider it a victory when they wake up in the morning and are still alive. Unfortunately, homelessness is on the rise everywhere in the United States. Organizations in several cities have developed music programs for the homeless, usually involving choirs since no instruments are needed, only voices. These organizations are aware that singing together creates bonds, eases the sense of isolation, and encourages a sense of community. The Dallas Street Choir has become one of the most visible.

Jonathan Palant conducts two choirs at the University of Texas, Dallas, as well as a church choir and a community choir. In 2014, a friend of his sent him a choral work called the *Street Requiem*, written in memory of people who had died while living on the streets. She wondered if he might like to perform it with Credo, his 150-member community choir. Palant thought it seemed a bit disingenuous to perform a work about street people with no homeless people involved. He had already been conducting an ad hoc choir for the homeless two or three times a year in connection with the Stewpot, a day shelter for the homeless in Dallas. He put up some signs, nine homeless individuals turned up at the first rehearsal, and the Dallas Street Choir was born. Before long, eighty to ninety individuals were coming to weekly rehearsal. Over 2,000 individuals have attended rehearsals since the founding of the choir.[15]

The Dallas Street Choir seeks to change the way society views the homeless through public performances at a variety of venues in the Dallas–Ft. Worth area. The fact that Palant thinks big has elevated not only the choir's visibility, but the visibility of the homeless. A longtime admirer of mezzo-soprano Frederica von Stade, he approached her about singing the soprano role in the premiere of the *Street Requiem*, the choral work by composers Kathleen McGuire, Andy Payne, and Jonathon Welch that spawned the idea of a choir for the homeless. To his surprise, she said yes, and she has been a supporter of the choir ever since. When in town performing with the Dallas Opera, she comes to rehearsals and works with the singers, and she has performed with them at Carnegie Hall. Yes, Carnegie Hall. Palant contacted Carnegie Hall to see if they had ever considered having a homeless choir perform, and in 2017, members of the Dallas Street Choir performed there, along with von Stade, soprano Harolyn Blackwell, and composers Jake Heggie and Stephen Schwartz.

Performing as a choir gives these homeless people a voice, and positive audience reactions build self-esteem. Performing in Carnegie Hall raised the profile of that voice. The singers felt a part of the larger world. The mantra of the Dallas Street Choir is "homeless, not voiceless," and members of the group want people to know that being homeless isn't a disability. Most members of the choir have high school diplomas, and many have college experience. Most had careers until something unexpected happened. Without families or estranged from them, they found themselves without a safety net and ended up on the street. They lost the sense of identity that comes with having a job. Many went from making important daily work decisions to facing basic questions of survival: "where will I sleep?" and "what will I eat?"

The Dallas Street Choir creates a safety net, a kind of "safe haven" for its homeless singers, a place they go once a week that provides consistency, structure, and accountability. They speak of the choir as family; they feel accepted and even valued; they share experiences (see Figure E.1 for a

Figure E.1 The Dallas Street Choir
Credit: Photo used by permission of Jonathan Palant

portrait of the ensemble). Some believe that singing in the choir opens new possibilities and helps them develop a capacity for change; others see the choir as a way to develop self-confidence, as a healing experience, a way to calm depression, and feel that music-making is an outlet to express emotions that can't be expressed on the street.[16] When people sync together musically, they are more likely to put the good of the group ahead of their own goals. One member of the choir commented: "I know I can do the little diva things sometimes, but then I realized, this is a group effort. It's not about me . . . it's about the choir working together in unity."[17]

After the performance at Carnegie Hall and a subsequent performance in Washington, DC, the choir returned to Dallas and the homeless members returned either to the streets or to shelters. Making music doesn't necessarily move adult homeless off the streets, but it gives participants a sense of community, self-esteem, and sometimes the self-confidence to explore new options. They may have lost hope today, but singing with the choir can give them hope for the future. *[Links to the Dallas Street Choir as well as other homeless choirs are found on the companion website at item E.3 ⊙.]*

Back at Harmony Project

We saw in Chapter 10 that students in Harmony Project showed brain changes positively affecting speech processing and language after studying an instrument for two years. But making music as part of a group also leads to pro-social behavior—trust of, and working together with, people who are different from you.

All kids need to feel connection with a group, and in the neighborhoods where Harmony Project kids live, gangs have typically provided that affiliation. But according to founder, Margaret Martin, Harmony Project creates a different kind of gang, an organization built on harmony and community rather than on crime and violence—a functional as opposed to a dysfunctional group. Harmony Project is primarily a mentoring program, and music is the vehicle to help kids learn to be responsible, accountable, competent, pro-social individuals. The children are in rehearsals or lessons for several hours a week, working with artist teachers/mentors but also with peer mentors, students like themselves who have been in the program longer and can offer guidance. During training to be a peer mentor, one young man said that when he learned he had been recommended by his Harmony Project teacher to be a peer mentor, he knew in that moment he would need to distance himself from his cousins who were members of a gang. If he was going to be a role model for other kids, he couldn't continue to affiliate with gang members.[18]

Harmony Project students come into the program with what Martin refers to as "the tyranny of low expectations." They may want to learn to play the instrument they have been given, but they don't really believe they will be able to do so. No one has ever believed in them before, not parents, teachers, or siblings. But the artist-teachers not only have expectations, they also believe the children can live up to those expectations. The young students surprise themselves by actually learning to play the instrument and making music. They develop self-confidence, as well as autonomy, accountability, self-reliance. Harmony Project becomes their refuge, a second family, and a sanctuary. It's a place they can go where all the other kids are there because they want to be—and they are all making music.

Bridging Differences

Martin relates that in Los Angeles, the parents of the kids in Harmony Project often come from rival gangs, gangs that kill each other. But the parents come together, sit in the audience, and applaud for all the kids. As they play in an orchestra, the students are all working together to perform a piece of music.

They come to understand that every part matters, including theirs. No matter what instrument has the melody, the complete piece doesn't exist without all the other instruments. Making music and sharing a common goal lead to better understanding of others as well as more confidence in oneself.

Jonathan Palant also speaks about bridging differences: "It doesn't matter if you are white, brown, black, it doesn't matter if you're Jewish, Christian, or Muslim, gay, straight, or trans. In the choral rehearsal, everybody's on an equal playing field and everybody's important in the rehearsal room. I think there's a huge sense of community because there is no hierarchy of any kind. Even though you are singing individually, everybody is singing the same thing: Do-mi-sol-mi-do. And for that reason, the community is really quite strong."[19]

Martin sees the kind of music-making that Harmony Project engages in as a way to heal our very divided nation. Others see music as a vehicle for bridging other kinds of divides. The West-Eastern Divan Orchestra was founded by Edward Said, Palestinian author and scholar, and pianist/conductor Daniel Barenboim as an alternative way to address the Israeli-Palestinian conflict. The base of the orchestra comprises an equal number of young Arab and Israeli musicians with additional members from Turkey, Iran, and Spain. In a 2005 speech after the orchestra played in Ramallah, Barenboim said, "Great music is the result of concentrated listening. Every musician listens intently to the voice of the composer and to each other. Harmony in personal or international relations can also only exist through listening, each party opening its ears to the other's narrative or point of view."[20]

The ancients knew that music builds community, as did Ernest Shackleford a hundred years ago. And we see it today everywhere—if we look. People need music in their lives. It is a way to express emotions and feel emotionally connected to others. It is a way to bridge divisions, share common goals, and create something together. Music provides a mechanism to understand others who are perceived as different. It is a crucial part of our deepest humanity.

Notes

Chapter 1

*. T. S. Eliot, "The Dry Salvages," No. 3 from *Four Quartets* (New York: Houghton Mifflin Harcourt, 1943), 44.

1. "Time with Tunes: How Technology is Driving Music Consumption," accessed February 3, 2020, https://www.nielsen.com/us/en/insights/article/2017/time-with-tunes-how-technology-is-driving-music-consumption/.

2. E. M. Hartwell, *Physical Training in American Colleges and Universities*, Circulars of Information of the Bureau of Education 5 (Washington, DC: US Government Printing Office, 1886), accessed January 15, 2020, https://archive.org/details/physicaltraining00hartiala/page/n4/mode/2up.

3. A. Pascual-Leone, A. Amedi, F. Fregni, and L. B. Merabet, "The Plastic Human Brain Cortex," *Annual Reviews Neuroscience* 28 (2005): 377–401.

4. G. Kochevitsky, *The Art of Piano Playing: A Scientific Approach* (New York: Summy-Birchard Music, 1967), 11.

5. R. Zatorre, "Music, the Food of Neuroscience?," *Nature* 434 (2005): 312–15.

6. L. M. Parsons, J. Sergent, D. A. Hodges, and P. T. Fox, "The Brain Basis of Piano Performance," *Neuropsychologia* 43, no. 2 (2005): 199–215.

7. C. J. Limb and A. R. Braun, "Neural Substrates of Spontaneous Musical Performance: An fMRI Study of Jazz Improvisation," *PLoS ONE* 3, no. 2 (2008): e1679, https://doi.org/10.1371/journal.pone.0001679.

8. "A Mind for Music," *NOVA*, Public Broadcasting Service, Harrisburg, PA WITF, June 29, 2009, https://www.pbs.org/wgbh/nova/video/a-mind-for-music/.

9. I. Peretz, "The Biological Foundations of Music: Insights from Congenital Amusia," in *The Psychology of Music*, 3rd ed., ed. D. Deutsch (Amsterdam: Elsevier Academic Press, 2013), 551–64.

10. M. Segelman, "Vissarion Shebalin," accessed April 27, 2021, https://www.wisemusicclassical.com/composer/2984/Vissarion-Shebalin/.

11. R. A. Henson, "Maurice Ravel's Illness: A Tragedy of Lost Creativity," *British Medical Journal* 296, no. 6636 (June 4, 1988): 1585–88; J. D. Warren and J. D. Rohrer, "Ravel's Last Illness: A Unifying Hypothesis," *Brain* 132 (2009): 1–2.

12. I. Peretz and D. T. Vuvan, "Prevalence of Congenital Amusia," *European Journal of Human Genetics* 25 (2017): 625–30; A. J. Sihvonen, T. Särkämö, P. Ripollés, V. Leo, J. Saunavaara, R. Parkkola, A. Rodríguez-Fornells, and S. Soinila, "Functional Neural Changes Associated with Acquired Amusia across Different Stages of Recovery after Stroke," *Scientific Reports* 7 (2017): 11390.

13. O. Sacks, *Musicophilia: Tales of Music and the Brain* (New York: Alfred A. Knopf, 2007).

14. O. Sacks, *The Mind's Eye* (New York: Alfred A. Knopf, 2010).
15. Ibid., 5–6.

Chapter 2

*. S. Mithen, "The Music Instinct: The Evolutionary Basis of Musicality," *Annals of the New York Academy of Sciences* 1169 (2009): 4

1. A. Brandt, M. Gebrian, and L. R. Slevc, "Music and Early Language Acquisition," *Frontiers in Psychology* 3 (2012), https://doi.org/10.3389/fpsyg.2012.00327.

2. N. J. Conard, M. Malina, and S. C. Münzel, "New Flutes Document the Earliest Musical Tradition in Southwestern Germany," *Nature* 460 (2009): 737–40.

3. Ibid.

4. T. Higham, L. Basell, R. Jacobi, R. Wood, C. B. Ramsey, and N. J. Conard, "Testing Models for the Beginnings of the Aurignacian and the Advent of Figurative Art and Music: The Radiocarbon Chronology of Gießenklösterle," *Journal of Human Evolution* 62 (2012): 664–76.

5. Conard, Malina, and Münzel, "New Flutes."

6. I. Morley, *The Prehistory of Music: Human Evolution, Archaeology, and the Origins of Musicality* (Oxford: Oxford University Press, 2013).

7. C. Darwin, *Descent of Man and Selection in Relation to Sex* (London: John Murray, 1871), 878.

8. W. James, *Principles of Psychology* (New York: Henry Holt, 1890), Retrieved from https://archive.org/details/theprinciplesofp01jameuoft/page/324/mode/2up?q= music, Vol. I:325, https://archive.org/details/theprinciplesofp00jameuoft/page/418/ mode/2up?q=nervous+system, Vol II:419, https://archive.org/details/theprincipl esofp00jameuoft/page/626/mode/2up?q=stairs, Vol II:627.

9. D. J. Levitin, *This Is Your Brain on Music: The Science of a Human Obsession* (New York: Dutton, 2006), 241–43.

10. S. Pinker, *How the Mind Works* (New York: W. W. Norton, 1997), 528–29.

11. S. Mithen, *The Singing Neanderthals: The Origins of Music, Language, Mind and Body* (Cambridge, MA: Harvard University Press, 2006), 109.

12. H. Honing, C. ten Cate, I. Peretz, and S. E. Trehub, "Without It No Music: Cognition, Biology and Evolution of Musicality," *Philosophical Transactions of the Royal Society B* 370 (2015), https://doi.org/10.1098/rstb.2014.0088.

13. I. Peretz, "The Nature of Music from a Biological Perspective," *Cognition* 100 (2006): 1–32; E. Bigand and B. Poulin-Charronnat, "Are We 'Experienced Listeners?' A Review of the Musical Capacities That Do Not Depend on Formal Musical Training," *Cognition* 100, no. 1 (2006): 100–30.

14. S. Dalla Bella and I. Peretz, "Differentiation of Classical Music Requires Little Learning but Rhythm," *Cognition* 96 (2005): B65–78.

15. Peretz, "Nature of Music," 4.

16. I. Peretz and R. Zatorre, eds., Preface to *The Cognitive Neuroscience of Music* (New York: Oxford University Press, 2003).

17. Mithen, *Singing Neanderthals.*

18. S. Mithen, "The Music Instinct: The Evolutionary Basis of Musicality," *Annals of the New York Academy of Sciences* 1169 (2009): 3–12.

19. N. L. Wallin, B. Merker, and S. Brown, eds., *The Origins of Music* (Cambridge: MIT Press, 2000).

20. Mithen, *Singing Neanderthals.*

21. R. D'Anastasio, S. Wroe, C. Tuniz, L. Mancini, D. T. Cesana, D. Dreossi, M. Ravichandiran, M. Attard, W. C. H. Parr, A. Agur, and L. Capasso, "Micro-Biomechanics of the Kebara 2 Hyoid and Its Implications for Speech in Neanderthals," *PLoS One* 8, no. 12 (2013): https://doi.org/10.1371/journal.pone.0082261.

22. Ibid.

23. Mithen, *Singing Neanderthals.*

24. Ibid.

25. L. C. Aiello, "Terrestriality, Bipedalism and the Origin of Language," in *Evolution of Social Behavior Patterns in Primates and Man*, ed. W. G. Runciman, J. Maynard-Smith, and R. I. M. Dunbar (Oxford: Oxford University Press, 1996), 269–90.

26. J. Montagu, "How Music and Instruments Began: A Brief Overview of the Origin and Entire Development of Music, from Its Earliest Stages," *Frontiers in Sociology* 2, no. 8 (June 20, 2017), https://doi.org/10.3389/fsoc.2017.00008.

27. S. N. Bibikov, "A Stone Age Orchestra: The Earliest Musical Instruments Were Made from the Bones of Mammoths," *The UNESCO Courier: A Window Open on the World* 28, no. 6 (1975): 28–31.

28. Ibid.

29. I. Reznikoff, "Sound Resonance in Prehistoric Times: A Study of Paleolithic Painted Caves and Rocks," *Journal of the Acoustical Society of America* 123, no. 5 (2008): 4137–41.

30. B. Fazenda, C. Scarre, R. Till, R. J. Pasalodos, M. R. Guerra, C. Tejedor, R. O. Peredo, A. Watson, S. Wyatt, C. G. Benito, H. Drinkall, and F. Foulds, "Cave Acoustics in Prehistory: Exploring the Association of Palaeolithic Visual Motifs and Acoustic Response," *Journal of the Acoustical Society of America* 142, no. 3 (2017), https://doi.org/10.1121/1.4998721.

31. "Q & A Rupert Till: Acoustic Archaeologist," interview with J. Hoffman, *Nature* 506 (2014): 158.

32. Ibid.

33. Reznikoff, "Sound Resonance."

34. J. Atema, "Musical Origins and the Stone Age Evolution of Flutes," *Acoustics Today* (Summer 2014): 26–34.

35. I. Turk, ed., "Mousterian Bone Flute and Other Finds from Divje Babe I Cave Site, Slovenia," Slovenian Institute for Archaeology (Ljubljana: Založba ZRC, 1997).

36. M. Turk and G. Bastiani, "Experimental Research on the Neanderthal Musical Instrument from Divje Babe I Cave (Slovenia)," *EXARC Journal* 2020, no. 3 (2020), https://exarc.net/ark:/88735/10522.

37. B. L. Hardy, M.-H. Moncel, C. Kerfant, M. Lebon, L. Bellot-Gurlet, and N. Mélard, "Direct Evidence of Neanderthal Fibre Technology and Its Cognitive and Behavioral Implications," *Scientific Reports* 10 (2020), https://doi.org/10.1038/s41

598-020-61839-w; D. L. Hoffman, C. D. Standish, M. Garcia-Diez, P. B. Pettitt, J. A. Milton, J. Zilhão, J. J Alcolea-Gonzálex, P. Cantalejo-Duarte, H. Collado, A. W. G. Pike, et al., "U-Th Dating of Carbonate Crusts Reveals Neandertal Origin of Iberian Cave Art," *Science* 359, no. 6378 (2018): 912–15; D. L. Hoffman, D. E. Angelucci, V. Villaverde, J. Zapata, and J. Zilhão, "Symbolic Use of Marine Shells and Mineral Pigments by Iberian Neandertals 115,000 Years Ago," *Science Advances* 4, no. 2 (2018), https://doi.org/10.1126/sciadv.aar5255; P. R. B. Kozowyk, M. Soressi, D. Pomstra, and G. H. J. Langejans, "Experimental Methods for the Palaeolithic Dry Distillation of Birch Bark: Implications for the Origin and Development of Neandertal Adhesive Technology," *Scientific Reports* 7 (2017), https://doi.org/10.1038/s41598-017-08106-7.

38. *Cave of Forgotten* Dreams, directed by W. Herzog (Creative Differences, 2010), 1:30, https://www.imdb.com/title/tt1664894/.

39. "'Useless' Art: Why Did a Prehistoric Artist Make This Sculpture?," BBC Civilisations, March 1, 2018, accessed May 20, 2020, https://www.bbc.co.uk/programmes/articles/59hXx0kKpv2mZKK3h6Z4dDl/useless-art-why-did-a-prehistoric-artist-make-this-sculpture.

40. A. Damasio, *Descartes' Error: Emotion, Reason, and the Human Brain* (New York: Putnam Publishing, 1994).

41. Mithen, "Music Instinct," 6.

Chapter 3

1. I. Winkler, G. P. Háden, O. Ladinig, I. Sziller, and H. Honing, "Newborn Infants Detect the Beat in Music," *Proceedings of the National Academy of Sciences* 106, no. 7 (2009): 2468–71.

2. A. Brandt, M. Gebrian, and L. R. Slevc, "Music and Early Language Acquisition," *Frontiers in Psychology* 3 (September 11, 2012), https://doi.org/10.3389/fpsyg.2012.00327.

3. H. Honing, C. ten Cate, I. Peretz, and S. E. Trehub, "Without It No Music: Cognition, Biology and Evolution of Musicality," *Philosophical Transactions B of the Royal Society* 370, no. 1664 (March 19, 2015), https://doi.org/10.1098/rstb.2014.0088.

4. M. Zentner and T. Eerola, "Rhythmic Engagement with Music in Infancy," *Proceedings of the National Academy of Sciences* 107, no. 13 (2010): 5768–73.

5. S. Kirschner and M. Tomasello, "Joint Drumming: Social Context Facilitates Synchronization in Preschool Children," *Journal of Experimental Child Psychology* 102 (2009): 299–314.

6. J. Phillips-Silver and L. J. Trainor, "Feeling the Beat: Movement Influences Infant Rhythm Perception," *Science* 308, no. 5727 (2005): 1430.

7. E. Hannon and S. Trehub, "Metrical Categories in Infancy and Adulthood," *Psychological Science* 16, no. 1 (2005): 48–55.

8. E. E. Hannon and S. E. Trehub, "Tuning In to Musical Rhythms: Infants Learn More Readily Than Adults," *Proceedings of the National Academy of Sciences* 102, no. 35 (August 30, 2005): 12639–643.

9. P. K. Kuhl, "Early Language Acquisition: Cracking the Speech Code," *Nature Reviews Neuroscience* 5 (2004): 831–43.

10. S. Mithen, *The Singing Neanderthals: The Origins of Music, Language, Mind, and Body.* (Cambridge, MA: Harvard University Press, 2006).

11. R. P. Cooper and R. N. Aslin, "Preference for Infant-Directed Speech in the First Month after Birth," *Child Development* 61 (1990): 1584–95.

12. M. Corbeil, S. E. Trehub, and I. Peretz, "Speech vs. Singing; Infants Choose Happier Sounds," *Frontiers in Psychology* (June 26, 2013), https://doi.org/10.3389/fpsyg.2013.00372.

13. M. Corbeil, S. E. Trehub, and I. Peretz, "Singing Delays the Onset of Infant Distress," *Infancy* 21, no. 3 (2016): 373–91.

14. R. Flom, D. A. Gentile, and A. D. Pick, "Infants' Discrimination of Happy and Sad Music," *Infant Behavior and Development* 31 (2008): 716–28.

15. Brandt, Gebrian, and Slevc, "Music."

16. J. Plantinga and L. J. Trainor, "Melody Recognition by Two-Month-Old Infants," *Journal of the Acoustical Society of America* 125, no. 2 (2009), https://doi.org/10.1121/1.3049583.

17. E. Partanen, T. Kujala, M. Tervaniemi, and M. Huotilainen, "Prenatal Music Exposure Induces Long-Term Neural Effects," *PLoS One* 8, no. 10 (October 30, 2013), https://doi.org/10.1371/journal.pone.0078946.

18. J. R. Saffran, M. M. Loman, and R. R. W. Robertson, "Infant Memory for Musical Experiences," *Cognition* 77 (2000): B15–23.

19. N. Kraus, E. Skoe, A. Parbery-Clark, and R. Ashley, "Experience-Induced Malleability in Neural Encoding of *Pitch, Timbre* and *Timing,*" *Annals of the New York Academy of Sciences* 1169 (2009): 543–57.

20. E. A. Piazza, M. C. Iordan, and C. Lew-Williams, "Mothers Consistently Alter Their Unique Vocal Fingerprints When Communicating with Infants," *Current Biology* 27 (2017): 3162–67.

21. L. J. Trainor, K. Lee, and D. J. Bosnyak, "Cortical Plasticity in 4-Month-Old Infants: Specific Effects of Experience with Musical Timbres," *Brain Topography* 24, no. 3–4 (2011): 192–203.

22. H. W. Chang and S. E. Trehub, "Auditory Processing of Relational Information by Young Infants," *Journal of Experimental Child Psychology* 24 (1977): 324–31.

23. S. E. Trehub and L. A. Thorpe, "Infants' Perception of Rhythm: Categorization of Auditory Sequences by Temporal Structure," *Canadian Journal of Psychology* 43, no. 2 (1989): 217–29.

24. J. Plantinga and S. E. Trehub, "Revisiting the Innate Preference for Consonance," *Journal of Experimental Psychology: Human Perception and Performance* 40 (2014): 40–49.

25. Brandt, Gebrian, and Slevc, "Music," 5.

26. Ibid.

27. A. J. Curwen, "Music Teaching," *Parents' Review* 9, no. 7 (1898): 439–62.

28. B. E. McPherson and A. Gabrielsson, "From Sound to Sign," in *The Science and Psychology of Music Performance,* ed. Richard Parncutt and Gary McPherson (Oxford: Oxford University Press, 2002), 99–115.

29. E. E. Gordon, *Learning Sequences in Music: A Contemporary Music Learning Theory, 2012 Edition* (Chicago: GIA Publications, 2012).

30. Brandt, Gebrian, and Slevc, "Music," 3.

31. E. E. Gordon, "A Music Learning Theory for Newborn and Young Children" (Chicago: GIA Publications, 1990), 2–3.

32. Gordon, *Learning Sequences.*

33. D. Gerry, A. Unrau, and L. J. Trainor, "Active Music Classes in Infancy Enhance Musical, Communicative and Social Development," *Developmental Science* 15 (2012): 398–407.

34. L. J. Trainor, C. Marie, D. Gerry, E. Whiskin, and A. Unrau, "Becoming Musically Enculturated: Effects of Music Classes for Infants on Brain and Behavior," *Annals of the New York Academy of Sciences* 1252 (April 2012): 129–38.

35. D. Barenboim, *Daniel Barenboim: A Life in Music*, ed. Michael Lewin (New York: Scribner, 1992); L. Barnett, "Portrait of the Artist: Daniel Barenboim, pianist." *The Guardian*, January 28, 2008, accessed May 1, 2021, https://www.theguardian.com/music/2008/jan/29/classicalmusicandopera.

36. A. Schein, personal communication, July 11, 2021.

Chapter 4

*. M. Merzenich, "How You Can Make Your Brain Smarter Every Day," *Forbes Next Avenue*, August 6, 2013, https://www.forbes.com/sites/nextavenue/2013/08/06/how-you-can-make-your-brain-smarter-every-day/?sh=422ddb5334ef.

1. W. James, "Habit" (New York: Henry Holt, 1914), accessed June 27, 2021, https://archive.org/details/habitjam00jameuoft/page/n6; W. James, *Principles of Psychology.* (New York: Henry Holt and Company, 1890).

2. S. Ramón y Cajal, *Degeneration & Regeneration of the Nervous System*, ed./trans. R. M. May (London: Oxford University Press, 1928; 1991).

3. W. Penfield and T. Rasmussen, *The Cerebral Cortex of Man* (New York: Macmillan, 1950).

4. H. Dempsey-Jones, D. B. Wesselink, J. Friedman, and T. R. Makin, "Organized Toe Maps in Extreme Foot Users," *Cell Reports* 28, no. 11 (2019): 2748–56.

5. M. M. Merzenich, R. J. Nelson, M. P. Stryker, M. S. Cynader, A. Schoppmann, and J. M. Zook, "Somatosensory Cortical Map Changes Following Digit Amputation in Adult Monkeys," *Journal of Comparative Neurology* 224 (1984): 591–605; S. A. Clark, T. Allard, W. M. Jenkins, and M. M. Merzenich, "Receptive Fields in the Body-Surface Map in Adult Cortex Defined by Temporally Correlated Inputs," *Nature* 332 (1988): 444–45; T. Allard, S. A. Clark, W. M. Jenkins, and M. M. Merzenich, "Reorganization of Somatosensory Area 3b in Representations in Adult Owl Monkeys after Digital Syndactyly," *Journal of Neurophysiology* 66 (1991): 1048–58.

6. A. Pascual-Leone and F. Torres, "Plasticity of the Sensorimotor Cortex Representation of the Reading Finger in Braille Readers," *Brain* 116, part 1 (1993): 39–52.

7. A. Pascual-Leone, D. Nguyet, L. G. Cohen, J. P. Brasil-Neto, A. Cammarota, and M. Hallett, "Modulation of Muscle Responses Evoked by Transcranial Magnetic Stimulation during the Acquisition of New Fine Motor Skills," *Journal of Neuropsychology* 74, no. 3 (1995): 1037–45.

8. A. Pascual-Leone, "The Brain That Plays Music and Is Changed by It," *Annals of the New York Academy of Sciences* 930 (2001): 315–29.

9. R. J. Zatorre, J. L. Chen, and V. B. Penhune, "When the Brain Plays Music: Auditory-Motor Interactions in Music Perception and Production," *Nature Reviews Neuroscience* 8 (2007): 547–58.

10. M. Bangert, U. Haeusler, and E. Altenmüller, "On Practice: How the Brain Connects Piano Keys and Piano Sounds," *Annals of the New York Academy of Sciences* 930, no. 1 (2001): 425–28.

11. J. Haueisen and T. R. Knösche, "Involuntary Motor Activity in Pianists Evoked by Music Perception," *Journal of Cognitive Neuroscience* 13, no. 6 (2001): 786–92.

12. M. Lotze, G. Scheler, H. R. M. Tan, C. Braun, and N. Birbaumer, "The Musician's Brain: Functional Imaging of Amateurs and Professionals during Performance and Imagery," *NeuroImage* 20 (2003): 1817–29.

13. M. Bangert and E. Altenmüller, "Mapping Perception to Action in Piano Practice: a Longitudinal DC-EEG Study," *BMC Neuroscience* (2003), http://www.biomedcentral. com/1471-2202/4/26.

14. *Music from the Inside Out*, DVD, directed by D. Anker (United States: Anker Productions, 2004).

15. E. Altenmüller and S. Furuya, "Apollos Gift and Curse: Making Music as a Model for Adaptive and Maladaptive Plasticity," *Neuroforum* 23, no. 2 (2017): A57–A75.

16. E. Altenmüller, "Music in Your Head," *Scientific American Mind* 14, no. 1 (2003): 24–31.

17. J. A. Grahn, "The Role of the Basal Ganglia in Beat Perception," *Annals of the New York Academy of Sciences* 1169 (2009): 35–45.

18. J. A. Grahn, "Neural Mechanisms of Rhythm Perception: Current Findings and Future Perspectives," *Topics in Cognitive Science* 4, no. 4 (2012): 585–606.

19. I. G. Meister, T. Krings, H. Foltys, B. Boroojerdi, M. Müller, R. Töpper, and A. Thron, "Playing Piano in the Mind—An fMRI Study on Music Imagery and Performance in Pianists," *Cognitive Brain Research* 19 (2004): 219–28.

20. C. Gaser and G. Schlaug, "Brain Structures Differ between Musicians and Non-Musicians," *Journal of Neuroscience* 23 (2003): 9240–45; L. Stewart, R. Henson, K. Kampe, V. Walsh, R. Turner, and U. Frith, "Brain Changes after Learning to Read and Play Music," *NeuroImage* 20 (2003): 71–83.

21. M. Koenigs, A. K. Barbey, B. R. Postle, and J. Grafman, "Superior Parietal Cortex Is Critical for the Manipulation of Information in Working Memory," *Journal of Neuroscience* 29, no. 47 (2009): 14980–86.

22. S. Hébert and L. L. Cuddy, "Music-Reading Deficiencies and the Brain," *Advances in Cognitive Psychology* 2, no. 2–3 (2006): 199–206.

23. F. A. C. Azevedo, "Equal Numbers of Neuronal and Nonneuronal Cells Make the Human Brain an Isometrically Scaled-Up Primate Brain," *Journal of Comparative Neurology* 513 (2009): 532–41.

24. D. O. Hebb, *The Organization of Behavior: A Neuropsychological Theory* (New York: Wiley, 1949).

25. I. S. Park, N. J. Lee, T.-Y. Kim, J.-H. Park, Y.-M. Won, Y.-J. Jung, J.-H. Yoon, and I. J. Rhyu, "Volumetric Analysis of Cerebellum in Short-Track Speed Skating Players," *Cerebellum* 11 (2012): 925–30; L. Jäncke, S. Koeneke, A. Hoppe, C. Rominger, and J. Hänggi, "The Architecture of the Golfer's Brain," *PLoS One* 4, no. 3 (2009), https://pubmed.ncbi.nlm.nih.gov/19277116/; E. A. Maguire, D. G. Gadian, I. S. Johnsrude, C. D. Good, J. Ashburner, R. S. J. Frackowiak, and C. D. Frith, "Navigation-Related Structural Change in the Hippocampi of Taxi Drivers," *Proceedings of the National Academy of Sciences* 97, no. 8 (2000): 4398–403; K. Woollett, H. J. Spiers, and E. A. Maguire, "Talent in the Taxi: A Model System for Exploring Expertise," *Philosophical Transactions of the Royal Society B* 364 (2009): 1407–16.

26. L. Jäncke, "Music Drives Brain Plasticity," *F1000 Biology Reports* 1, no. 10 (2009): 78.

27. L. Jäncke, "The Plastic Human Brain," *Restorative Neurology and Neuroscience* 27 (2009): 521–38.

28. N. Gaab, C. Gaser, and G. Schlaug, "Improvement-Related Functional Plasticity Following Pitch Memory Training," *NeuroImage* 31 (2006): 255–63; F. Talamini, G. Altoè, B. Carretti, and M. Grassi, "Musicians Have Better Memory Than Nonmusicians: A Meta-Analysis," *PLoS One* (2017), https://doi.org/10.1371/journal.pone.0186773; G. Schlaug, "Musicians and Music Making as a Model for the Study of Brain Plasticity," *Progress in Brain Research* 217 (2015): 37–55.

29. T. F. Münte, E. Altenmüller, and L. Jäncke, "The Musician's Brain as a Model of Neuroplasticity," *Nature Reviews Neuroscience* 3 (2002): 473–78.

30. T. Elbert, C. Pantev, C. Wienbruch, B. Rockstroh, and E. Taub, "Increased Cortical Representation of the Fingers of the Left Hand in String Players," *Science* 270, no. 5234 (1995): 305–7.

31. M. Bangert and G. Schlaug, "Specialization of the Specialized in Features of External Human Brain Morphology," *European Journal of Neuroscience* 26, no. 6 (2006): 1832–34.

32. Gaser and Schlaug, "Brain Structures."

33. V. Sluming, J. Brooks, M. Howard, J. J. Downes, and N. Roberts, "Broca's Area Supports Enhanced Visuospatial Cognition in Orchestral Musicians," *Journal of Neuroscience* 27, no. 14 (2007): 3799–806; L. M. Parsons, J. Sergent, D. A. Hodges, and P. T. Fox, "The Brain Basis of Piano Performance," *Neuropsychologia* 43 (2005): 199–215; S. L. Bengtsson and F. Ullén, "Dissociation between Melodic and Rhythmic Processing during Piano Performance from Musical Scores," *NeuroImage* 30 (2006): 272–84.

34. G. Schlaug, L. Jäncke, Y. Huang, J. F. Staiger, and H. Steinmetz, "Increased Corpus Callosum Size in Musicians," *Neuropsychologia* 33, no. 8 (1995): 1047–55; Y. Han, H. Yang, Y.-T. Lv, C.-Z. Zhu, Y. He, H.-H. Tang, Q.-Y. Gong, Y.-J. Luo, Y.-F. Zang, and Q. Dong, "Gray Matter Density and White Matter Integrity in Pianists' Brain: A Combined Structural and Diffusion Tensor MRI Study," *Neuroscience Letters* 459, no. 1 (2009): 3–6; S. Hutchinson, L. H. Lee, N. Gaab, and G. Schlaug, "Cerebellar Volume of Musicians," *Cerebral Cortex* 13 (2003): 943–49.

35. M. Groussard, R. La Joie, G. Rauchs, B. Landeau, G. Chételat, F. Viader, B. Desgranges, F. Eustache, and H. Platel, "When Music and Long-Term Memory Interact: Effects of

Musical Expertise on Functional and Structural Plasticity in the Hippocampus," *PLoS One* (October 5, 2010), https://journals.plos.org/plosone/article?id=10.1371/journal.pone.0013225; V. S. Sluming, D. Page, J. Downe, D. Denby, A. Mayes, and N. Roberts, "Increased Hippocampal Volumes and Enhanced Visual Memory in Musicians," *Proceedings of the International Society for Magnetic Resonance in Medicine* 13 (2005): 83.

36. B. Kleber, R. Veit, N. Birbaumer, J. Gruzelier, and M. Lotze, "The Brain of Opera Singers: Experience-Dependent Changes in Functional Activation," *Cerebral Cortex* 20 (2010): 1144–52.

37. Schlaug et al., "Increased Corpus Callosum"; K. L. Hyde, J. Lerch, A. Norton, M. Forgeard, E. Winner, A. C. Evans, and G. Schlaug, "Musical Training Shapes Structural Brain Development," *Journal of Neuroscience* 29 (2009): 3019–25.

38. S. L. Bengtsson, A. Nagy, S. Skare, L. Forsman, H. Forssberg, and F. Ullén, "Extensive Piano Practicing Has Regionally Specific Effects on White Matter Development," *Nature Neuroscience* 8, no. 9 (2005): 1148–50.

39. T. Fujioka, B. Ross, R. Kakigi, C. Pantev, and L. J. Tranior, "One Year of Musical Training Affects Development of Auditory Cortical-Evoked Fields in Young Children," *Brain* 129, part 10 (2006): 2593–608; Hyde et al., "Musical Training"; G. Schlaug, M. Forgeard, L. Zhu, A. Norton, A. Norton, and E. Winner, "Training-Induced Neuroplasticity in Young Children," *Annals of the New York Academy of Sciences* 1169 (2009): 205–8; A. Habibi, A. Damasio, B. Ilari, R. Veiga, A. A. Joshi, R. M. Leahy, J. P. Haldar, D. Varadarajan, C. Bhushan, and H. Damasio, "Childhood Music Training Induces Change in Micro and Macroscopic Brain Structure: Results from a Longitudinal Study," *Cerebral Cortex* 28, no. 12 (2018): 4336–47.

40. L. Jäncke, N. J. Shah, and M. Peters, "Cortical Activations in Primary and Secondary Motor Areas for Complex Bimanual Movements in Professional Pianists," *Cognitive Brain Research* 10 (2000): 177–83.

41. O. Granert, M. Peller, H.-C. Jabusch, E. Altenmüller, and H. R. Siebner, "Sensorimotor Skills and Focal Dystonia are Linked to Putaminal Grey-Matter Volume in Pianists," *Journal of Neurology, Neurosurgery and Psychiatry* 82 (2011): 1225–31; E. E. James, M. S. Oechslin, D. Van de Ville, C.-A. Hauert, C. Descloux, and F. Lazeyras, "Musical Training Intensity Yields Opposite Effects on Grey Matter Density in Cognitive versus Sensorimotor Networks," *Brain Structure and Function* 219 (2014): 353–66.

42. N. Kraus and T. White-Schwoch, "Unraveling the Biology of Auditory Learning: A Cognitive-Sensorimotor-Reward Framework," *Trends in Cognitive Sciences* 19, no. 11 (2015): 642–54.

Chapter 5

*. J. Austen, *Mansfield Park* (London: Penguin Classics, 1966; orig. London: Thomas Egerton, 1814), 222.

1. A. Pascual-Leone, "The Brain That Plays Music and Is Changed by It," *Annals of the New York Academy of Sciences* 930 (2001): 315–29.

2. J. A. Kelso, "Phase Transitions and Critical Behavior in Human Bimanual Coordination," *American Journal of Physiology* 246 (1984): R1000–4; S. P. Swinnen, "Intermanual Coordination: From Behavioural Principles to Neural-Network Interactions," *Nature Reviews Neuroscience* 3 (2002): 348–59.

3. S. P. Swinnen and N. Wenderoth, "Two Hands, One Brain: Cognitive Neuroscience of Bimanual Skill," *TRENDS in Cognitive Sciences* 8, no. 1 (2004): 18–25.

4. J. LeDoux, *Synaptic Self: How Our Brains Become Who We Are* (New York: Penguin Books, 2002), 161.

5. E. A. Phelps, "Human Emotion and Memory: Interactions of the Amygdala and Hippocampal Complex," *Current Opinion in Neurobiology* 12 (2004): 198–202; S. B. Hamann, T. D. Ely, J. M. Hoffman, and C. D. Kilts, "Ecstasy and Agony: Activation of the Human Amygdala in Positive and Negative Emotion," *Psychological Science* 13, no. 2 (2002): 135–41.

6. D. Levitin, *This Is Your Brain on Music: The Science of a Human Obsession* (New York: Dutton, 2006), 225.

7. E. A. Kensinger, "Remembering the Details: Effects of Emotion," *Emotion Review* 1, no. 2 (2009): 99–113.

8. J. J. Nagel, *Managing Stage Fright: A Guide for Musicians and Music Teachers* (New York: Oxford University Press, 2017); V. Cornett, *The Mindful Musician: Mental Skills for Peak Performance* (New York: Oxford University Press, 2019).

9. E. Landau, "It's in Your Head, Changing Your Brain," posted May 28, 2012, http://www.cnn.com/2012/05/26/health/mental-health/music-brain-science/.

10. L. Boyle, "Now This Masterpiece COULD Drive You Mad! Domino Artist Takes a Painstaking 11 Hours to Recreate van Gogh's Starry Night (and It's Gone in Just Seconds)," *DailyMail.com*, posted June 25, 2012, https://www.dailymail.co.uk/news/article-2164727/Van-Goghs-Starry-Night-recreated-domino-artist-FlippyCat.html.

11. R. Chaffin and G. Imreh, "Practicing Perfection: Piano Performance as Expert Memory," *Psychological Science* 13 (2002): 342–49; J. Ginsborg, R. Chaffin, and G. Nicholson, "Shared Performance Cues in Singing and Conducting: A Content Analysis of Talk during Practice," *Music Psychology* 34 (2006): 167–94; H. Noice, R. Chaffin, J. Jeffrey, and A. Noice, "Memorization by a Jazz Pianist: A Case Study," *Psychology of Music* 36 (2008): 47–61; R. Chaffin, T. R. Logan, and K. T. Begosh, "Performing from Memory," in *The Oxford Handbook of Music Psychology*, ed. S. Hallam, I. Cross, and M. Thaut (Oxford: Oxford University Press, 2009), 352–63.

12. Chaffin and Imreh, "Practicing Perfection."

Chapter 6

*. W. Westney, *The Perfect Wrong Note: Learning to Trust Your Musical Self* (Pompton Plains, NJ; Cambridge: Amadeus Press, 2003), 49.

1. R. A. Bjork and E. L. Bjork, "A New Theory of Disuse and an Old Theory of Stimulus Fluctuation," in *From Learning Processes to Cognitive Processes: Essays in Honor*

of William K. Estes, Vol. 2, ed. A. Healy, S. Kosslyn, and R. Shiffrin (Hillsdale, NJ: Erlbaum, 1992), 35–67.

2. R. A. Bjork, "Memory and Metamemory Considerations in the Training of Human Beings," in *Metacognition: Knowing about Knowing*, ed. J. Metcalfe and A. Shimamura (Cambridge, MA: MIT Press, 1994), 185–205; E. L. Bjork and R. A. Bjork, "Making Things Hard on Yourself, but in a Good Way: Creating Desirable Difficulties to Enhance Learning," in *Psychology and the Real World: Essays Illustrating Fundamental Contributions to Society*, ed. M. A. Gernsbacher, R. W. Pew, L. M. Hough, and J. R. Pomerantz (New York: Worth Publishers, 2011), 56–64.

3. J. Dunlosky, K. A. Rawson, E. J. Marsh, M. J. Nathan, and D. T. Willingham, "Improving Students' Learning with Effective Learning Techniques: Promising Directions from Cognitive and Educational Psychology," *Psychological Science in the Public Interest* 14, no. 1 (2013): 4–58.

4. N. Barry, "A Qualitative Study of Applied Music Lessons and Subsequent Student Practice Sessions," *Contributions to Music Education* 34 (2007): 51–65; D. Rohwer and J. Polk, "Practice Behaviors of Eighth-Grade Instrumental Musicians," *Journal of Research in Music Education* 54, no. 4 (2006): 350–62.

5. H. Ebbinghaus, *Memory: A Contribution to Experimental Psychology*, trans. H. A. Ruger and C. E. Bussenius (New York: Dover, 1964; orig. 1885).

6. A. D. Baddeley and D. J. A. Longman, "The Influence of Length and Frequency of Training Session on the Rate of Learning to Type," *Ergonomics* 21, no. 8 (1978): 627–35.

7. F. Brady, "The Contextual Interference Effect and Sport Skills," *Perceptual Motor Skills* 106 (2008): 461–72.

8. W. F. Battig, "Facilitation and Interference," in *Acquisition of Skill*, ed. E. A. Bilodeau (New York: Academic Press, 1966), 215–44.

9. J. B. Shea and R. L. Morgan, "Contextual Interference Effects on the Acquisition, Retention, and Transfer of a Motor Skill," *Journal of Experimental Psychology* 5 (1979): 179–87; J. M. Porter, D. Landin, E. P. Hebert, and B. Baum, "The Effects of Three Levels of Contextual Interference on Performance Outcomes and Movement Patterns in Golf Skills," *International Journal of Sports Science & Coaching* 22 (2007): 243–55; T. Buszard, M. Reid, L. Krause, S. Kovalchik, and D. Farrow, "Quantifying Contextual Interference and Its Effect on Skill Transfer in Skilled Youth Tennis Players," *Frontiers in Psychology* 8 (2017), https://doi.org/10.3389/fpsyg.2017.01931; P. J. K. Smith, "Applying Contextual Interference to Snowboarding Skills," *Perceptual and Motor Skills* 95 (2002): 999–1005.

10. N. L. Foster, M. L. Mueller, C. Was, K. A. Rawson, and J. Dunlosky, "Why Does Interleaving Improve Math Learning?: The Contributions of Discriminative Contrast and Distributed Practice," *Memory & Cognition* 47 (2019): 1088–1101; T. Linderholm, J. Dobson, and M. B. Yarbrough, "The Benefit of Self-Testing and Interleaving for Synthesizing Concepts Across Multiple Physiology Texts," *Advances in Physiology Education* 40 (2016): 329–34; S. K. Carpenter and F. E. Mueller, "The Effects of Interleaving versus Blocking on Foreign Language Pronunciation Learning," *Memory & Cognition* 41, no. 5 (2016): 671–82.

11. E. E. Carter and J. A. Grahn, "Optimizing Music Learning: Exploring How Blocked and Interleaved Practice Schedules Affect Advanced Performance," *Frontiers in Psychology* 7 (2016), https://doi.org/10.3389/fpsyg.2016.01251.

12. T. D. Lee and D. A. Simon, "Contextual Interference," in *Skill Acquisition in Sport: Research, Theory, Practice*, ed. A. M. Williams and N. J. Hodges (London: Routledge, 2004), 29–44.

13. R. A. Magill and K. G. Hall, "A Review of the Contextual Interference Effect in Motor Skill Acquisition," *Human Movement Science* 9 (1990): 241–89.

14. J. LeDoux, *Synaptic Self: How Our Brains Become Who We Are* (New York: Penguin Books, 2002), 106.

15. H. Pashler, M. McDaniel, D. Rohrer, and R. Bjork, "Learning Styles: Concepts and Evidence," *Psychological Science in the Public Interest* 9, no. 3 (2009): 105–19.

16. R. P. Shockley, *Mapping Music: For Faster Learning and Secure Memory: A Guide for Piano Teachers and Students* (Middleton, WI: A-R Editions, 1997).

17. J. D. Wammes, M. E. Meade, and M. A. Fernandes, "The Drawing Effect: Evidence for Reliable and Robust Memory Benefits in Free Recall," *Quarterly Journal of Experimental Psychology* 69, no. 9 (2016): 1752–76.

18. A. Whiteside, *Abby Whiteside on Piano Playing: Indispensables of Piano Playing— Mastering the Chopin Etudes and Other Essays* (Original essays, New York: Charles Scribner's Sons, 1955 and 1969; repr., Amadeus Press, 1997, 2003).

19. K. S. Lashley, "The Problem of Serial Order in Behavior," in *Cerebral Mechanisms in Behavior*, ed. L. A. Jeffress (New York: Wiley, 1951), 112–31.

20. E. Altenmüller and W. Gruhn, "Brain Mechanisms," in *The Science and Psychology of Music Performance: Creative Strategies for Teaching and Learning*, ed. R. Parncutt and G. E. McPherson (New York: Oxford University Press, 2002), 63–81.

21. E. Altenmüller, personal communication, July 1, 2021.

22. S. M. Smith, "Background Music and Context-Dependent Memory," *American Journal of Psychology* 98, no. 4 (1985): 591–603.

23. J. E. Eich, "State-Dependent Accessibility of Retrieval Cues in the Retention of a Categorized List," *Journal of Verbal Learning and Verbal Behavior* 14 (1975): 408–17.

24. S. M. Smith, A. Glenberg, and R. A. Bjork, "Environmental Context and Human Memory," *Memory & Cognition* 6, no. 4 (1978): 342–53.

25. L. Eustis, personal communication, May 26, 2021.

26. J. Steinbeck, *Sweet Thursday* (New York: Viking Press, 1954), 88.

27. J. G. Jenkins and K. M. Dallenbach, "Obliviscence during Sleep and Waking," *American Journal of Psychology* 35, no. 4 (1924): 605–12.

28. M. P. Walker, *Why We Sleep: Unlocking the Power of Sleep and Dreams* (New York: Scribner, 2017); M. P. Walker and R. Stickgold, "Sleep-Dependent Learning and Memory Consolidation," *Neuron* 44 (2004): 121–33.

29. M. P. Walker, "A Refined Model of Sleep and the Time Course of Memory Formation," *Behavioral and Brain Sciences* 28 (2005): 51–104.

30. Walker and Stickgold, "Sleep-Dependent Learning."

31. M. P. Walker, T. Brakefield, A. Morgan, J. A. Hobson, and R. Stickgold, "Practice with Sleep Makes Perfect: Sleep-Dependent Motor Skill Learning," *Neuron* 35

(2002): 205–11; M. P. Walker, R. Stickgold, D. Alsop, N. Gaab, and G. Schlaug, "Sleep-Dependent Motor Memory Plasticity in the Human Brain," *Neuroscience* 133, no. 4 (2005): 911–17; J. Doyon, M. Korman, A. Morin, V. Dostie, A. H. Tahar, H. Benali, A. Karni, L. G. Ungerleider, and J. Carrier, "Contribution of Night and Day Sleep vs. Simple Passage of Time to the Consolidation of Motor Sequence and Visuomotor Adaptation Learning," *Experimental Brain Research* 195 (2009): 15–26.

32. S. Allen, "Memory Stabilization and Enhancement Following Music Practice," *Psychology of Music* 41, no. 6 (2013): 691–712.

33. C. Smith and C. MacNeill, "Impaired Motor Memory for a Pursuit Rotor Task Following Stage 2 Sleep Loss in College Students," *Journal of Sleep Research* 3 (1994): 206–13.

34. S. P. Drummond and G. G. Brown, "The Effects of Total Sleep Deprivation on Cerebral Responses to Cognitive Performance," *Neuropsychopharmacology* 25, 5 Suppl (2001): S68–S73; S. P. Drummond, G. G. Brown, J. C. Gillin, J. L. Stricker, E. C. Wong, and R. B. Buxton, "Altered Brain Response to Verbal Learning Following Sleep Deprivation," *Nature* 403 (2000): 655–57.

35. E. B. Simon, A. Rossi, A. G. Harvey, and M. P. Walker, "Overanxious and Underslept," *Nature Human Behavior* 4 (2019): 100–10.

36. J. Cedernaes, F. H. R. Rångtell, E. K. Axelsson, A. Yeganeh, H. Vogel, J.-E. Broman, S. L. Dickson, H. B. Schiöth, and C. Benedict, "Short Sleep Makes Declarative Memories Vulnerable to Stress in Humans," *Sleep* 38, no. 12 (2015): 1861–68.

37. M. Nishida and M. P. Walker, "Daytime Naps, Motor Memory Consolidation and Regionally Specific Sleep Spindles," *PLoS One* 2, no. 4 (2007), https://doi.org/10.1371/journal.pone.0000341.

38. M. Gladwell, *Outliers: The Story of Success* (New York: Little, Brown & Co., 2008), 41.

39. K. A. Ericsson, R. T. Krampe, and C. Tesch-Römer, "The Role of Deliberate Practice in the Acquisition of Expert Performance," *Psychological Review* 100 (1993): 363–406.

40. A. Schein, personal communication, July 11, 2021.

Chapter 7

1. A. Warraich and J. A. Kleim, "Neural Plasticity: The Biological Substrate for Neurorehabilitation," *PM&R* 2 (2010): S208–S219.

2. A. H. Rutkin, "Champagne for the Blind: Paul Bach-y-Rita, Neuroscience's Forgotten Genius" (thesis, MIT, 2013), https://cmsw.mit.edu/paul-bach-y-rita-neurosciences-forgotten-genius/.

3. L. B. Merabet and A. Pascual-Leone, "Neural Reorganization Following Sensory Loss: The Opportunity of Change," *Nature Reviews Neuroscience* 11 (2010): 44–52.

4. Ibid.; A. Pascual-Leone and F. Torres, "Plasticity of the Sensorimotor Cortex Representation of the Reading Finger in Braille Readers," *Brain* 116 (1993): 39–52; T. Elbert, A. Sterr, B. Rockstroh, C. Pantev, M. M. Müller, and E. Taub, "Expansion of the Tonotopic Area in the Auditory Cortex of the Blind," *Journal of Neuroscience* 22 (2002): 9941–44.

5. A. Amedi, L. B. Merabet, J. Camprodon, F. Bermpohl, S. Fox, I. Ronen, D. S. Kim, and A. Pascual-Leone, "Neural and Behavioral Correlates of Drawing in an Early Blind Painter: A Case Study," *Brain Research* 1242 (2008): 252–62.

6. A. Amedi, K. von Kriegstein, N. M. van Atteveldt, M. S. Beauchamp, and M. J. Naumer, "Functional Imaging of Human Crossmodal Identification and Object Recognition," *Experimental Brain Research* 166 (2005): 559–71.

7. L. Reich, M. Szwed, L. Cohen, and A. Amedi, "A Ventral Visual Stream Reading Center Independent of Visual Experience," *Current Biology* 21, no. 5 (2011): 363–68.

8. E. Blake, "Blind Pianist Nobuyuki Tsujii at the Opera House with the Sydney Symphony Orchestra," *Sydney Morning Herald*, April 25, 2017, https://www.smh.com.au/entertainment/music/blind-pianist-nobuyuki-tsujii-at-the-opera-house-with-the-sydney-symphony-orchestra-20170425-gvs02c.html.

9. Y. Oda, "'Nobu' Fever: Japan Falls for a Blind Piano Prodigy," *Time*, November 18, 2009, http://content.time.com/time/world/article/0,8599,1940215,00.html; I. Warden, "Blind Japanese Virtuoso to Tickle Parliament House Ivories for PM," *Canberra Times*, October 13, 2016, https://www.canberratimes.com.au/story/6041341/blind-japanese-virtuoso-to-tickle-parliament-house-ivories-for-pm/.

10. M. Schafter, "Meet Nobuyuki Tsujii, the Blind Concert Pianist Who Learns by Ear," *Australian Broadcasting Corporation*, May 31, 2017, https://www.abc.net.au/7.30/meet-nobuyuki-tsujii,-the-blind-concert-pianist/8577524.

11. S. Nakajima, "Nobuyuki Tsujii: Vibration as Connection," *Parasapo*, Issue 3, Column 1, https://www.parasapo.tokyo/gojournal/en/issue03/column01.

12. D. Polkow, "Blind 'Miracle' Pianist a Marvel, yet Artistically a Work in Progress," *Chicago Classical Review*, June 4, 2010, https://chicagoclassicalreview.com/2010/06/blind-miracle-pianist-a-marvel-yet-artistically-a-work-in-progress.

13. I. Hewett, "Nobuyuki Tsujii: 'The Piano Is an Extension of My Own Body,'" *The Telegraph*, July 16, 2013, https://www.telegraph.co.uk/culture/music/proms/10182751/Nobuyuki-Tsujii-The-piano-is-an-extension-of-my-own-body.html.

14. *Touching the Sound: The Improbable Journey of Nobuyuki Tsujii*, directed by Peter Rosen (Peter Rosen Productions, 2015), 1:08, imdb.com/title/tt3559558.

15. A. Froug, "Q & A with Nobuyuki Tsujii: World-Renowned Pianist Shares His Thoughts on Art of Piano Performance," *Daily Bruin*, April 1, 2011, https://dailybruin.com/2011/04/01/qampa_with_nobuyuki_tsujii_worldrenowned_pianist_shares_his_thoughts_on_art_of_piano_performance.

16. S. Levänen and D. Hamdorf, "Feeling Vibrations: Enhanced Tactile Sensitivity in Congenitally Deaf Humans," *Neuroscience Letters* 301, no. 1 (2001): 75–77; D. Bavelier, A. Tomann, C. Hutton, T. Mitchell, D. Corina, G. Liu, and H. Neville, "Visual Attention to the Periphery Is Enhanced in Congenitally Deaf Individuals," *Journal of Neuroscience* 20, no. 17 (2000): RC93; S. McCullough and K. Emmorey, "Face Processing by Deaf ASL Signers: Evidence for Expertise in Distinguished Local Features," *Journal of Deaf Studies and Deaf Education* 2, no. 4 (1997): 212–22.

17. Merabet and Pascual-Leone, "Neural Reorganization"; C. M. Karns, M. W. Dow, and H. J. Neville, "Altered Cross-Modal Processing in the Primary Auditory Cortex of Congenitally Deaf Adults: A Visual-Somatosensory fMRI Study with a Double-Flash Illusion," *Journal of Neuroscience* 32, no. 28 (2012): 9626–38.

18. University of Washington, "Brains of Deaf People Rewire to 'Hear' Music," news release in ScienceDaily, November 28, 2001, www.sciencedaily.com/releases/2001/11/011128035455.htm.

19. E. Glennie, "Hearing Essay," January 1, 2015, https://www.evelyn.co.uk/hearing-essay/.

20. E. Glennie, *Good Vibrations* (London: Arrow Books, 1990).

21. E. Glennie, "I Have Seen and Touched the Sound," March 6, 2012, https://www.evelyn.co.uk/i-have-seen-and-touched-the-sound/.

22. E. Glennie, "How to Truly Listen," filmed February 2003, TED video, 31:56, https://www.ted.com/talks/evelyn_glennie_how_to_truly_listen?language=en.

23. E. Glennie, "Deaf, Sound and Music Questions," April 2019, https://www.evelyn.co.uk/wp-content/uploads/2019/06/Evelyn-Glennie-Deaf-and-Music-Questions.pdf.

24. *Touch the Sound: A Sound Journey with Evelyn Glennie*, DVD, directed by Thomas Riedelsheimer, November 4, 2004, Germany, UK: Filmquadrat/Skyline Productions, 2004.

25. E. Altenmüller and H-C. Jabusch, "Focal Dystonia in Musicians: Phenomenology, Pathophysiology, Triggering Factors, and Treatment," *Medical Problems of Performing Artists* 25, no. 1 (2010): 3–9.

26. Because so little information is available, singers' dystonia won't be discussed in this chapter; for information about case histories of four singers with focal laryngeal dystonia, see L. A. Halstead, D. M. McBroom, and H. S. Bonilha, "Task-Specific Singing Dystonia: Vocal Instability That Technique Cannot Fix," *Journal of Voice* 29, no. 1 (2015): 71–78.

27. R. E. Newby, D. E. Thorpe, P. A. Kempster, and J. E. Alty, "A History of Dystonia: Ancient to Modern," *Movement Disorders Clinical Practice* (March 30, 2017), https://doi.org/10.1002/mdc3.12493.

28. L. Fleisher and A. Midgette, *My Nine Lives: A Memoir of Many Careers in Music* (New York: Anchor Books, 2011).

29. A. Steinbach, "'The Gods Hit You Where It Hurts': Mysterious Hand Ailments Ambushed the Concert Careers of Leon Fleisher and Gary Graffman. How Each Met the Challenge Is Testament to Their Strength and Spirit. And to Their Friendship," *Baltimore Sun*, April 7, 1996, https://www.baltimoresun.com/news/bs-xpm-1996-04-07-1996098184-story.html.

30. C. D. Marsden and M. P. Sheehy, "Writer's Cramp," *Trends in Neurosciences* 13, no. 4 (1990): 148–153.

31. N. Byl, "Focal Hand Dystonia: A Historical Perspective from a Clinician Scholar," *Journal of Hand Therapy* 22, no. 2 (2009): 105–8.

32. T. Elbert, V. Candia, E. Altenmüller, H. Rau, A. Sterr, B. Rockstroh, C. Pantev, and E. Taub, "Alteration of Digital Representations in Somatosensory Cortex in Focal Hand Dystonia," *Neuroreport* 9, no. 16 (1998): 3571–75.

33. Y. Hirata, M. Schulz, E. Altenmüller, T. Elbert, and C. Pantev, "Sensory Mapping of Lip Representation in Brass Musicians with Embouchure Dystonia," *Neuroreport* 15, no. 5 (2004): 815–18.

34. V. Candia, C. Weinbruch, T. Elbert, B. Rockstroh, and W. Ray, "Effective Behavioral Treatment of Focal Hand Dystonia in Musicians Alters Somatosensory Cortical Organization," *Proceedings of the National Academy of Sciences* 100, no. 13 (2003): 7942–46.

35. D. Vining, comp., *Notes of Hope: Stories by Musicians Coping with Injuries* (Flagstaff, AZ: Mountain Peak Music, 2014), 42.

36. Elbert, "Alteration of Digital Representations," 3573; Hirata, "Sensory Mapping," 817.

37. E. Altenmüller, and S. Furuya, "Apollos Gift and Curse: Making Music as a Model for Adaptive and Maladaptive Plasticity," *Neuroforum* 23, no. 2 (2017): A57–A75.

38. E. Altenmüller, personal communication, July 1, 2021.

39. E. Altenmüller, "Focal Dystonia: Advances in Brain Imaging and Understanding of Fine Motor Control in Musicians," *Hand Clinics* 19 (2003): 523–38; Altenmüller and Furuya, "Apollos Gift", A67; E. Altenmüller, personal communication, July 1, 2021.

40. A. Schmidt, H. -C. Jabusch, E. Altenmüller, J. Hagenah, N. Brüggeman, K. Lohmann, L. Enders, P. L. Kramer, R. Saunders-Pullman, S. B. Bressman, A. Münchau, and C. Klein, "Etiology of Musician's Dystonia: Familial or Environmental?," *Neurology* 72, no. 14 (2009): 1248–54.

41. H. C. Jabusch, D. Zschucke, A. Schmidt, S. Schuele, and E. Altenmüller, "Focal Dystonia in Musicians: Treatment Strategies and Long-Term Outcome in 144 Patients," *Movement Disorders* 220, no. 12 (2005): 1623–26.

42. E. Altenmüller, personal communication, July 1, 2021.

43. C. Lungu, B. I. Karp, K. Alter, R. Zolbrod, and M. Hallett, "Long Term Follow-up of Botulinum Toxin Therapy for Focal Hand Dystonia: Outcome at 10 or More Years," *Movement Disorders* 26, no. 4 (2011): 750–53.

44. E. Altenmüller, personal communication, July 1, 2021.

45. E. Altenmüller, "The End of the Song?: Robert Schumann's Focal Dystonia," in *Music, Motor Control and the Brain*, ed. E. Altenmüller, M. Wiesendanger, and J. Kesselring (New York: Oxford University Press), 251–63.

46. B. Ackermann and E. Altenmüller, "The Development and Use of an Anatomy-Based Retraining Program (MusAARP) to Assess and Treat Focal Hand Dystonia in Musicians—A Pilot Study," *Journal of Hand Therapy* 34 (2021): 309–14.

47. F. Wilson, "Glenn Gould's Hand," in *Medical Problems of the Instrumentalist Musician*, ed. R. Tubiana and P. Amadio (London: Martin Dunitz, Ltd., 2000), 379–97.

48. J. Heggie, personal communication, April 9, 2021.

49. N. Skolnik, personal communication, August 21, 2021.

50. J. Heggie, personal communication, April 9, 2021.

51. J. Litzelman, personal communication, June 30, 2021.

52. S. Furuya, M. A. Nitsche, W. Paulus, and E. Altenmüller, "Surmounting Retraining Limits in Musicians' Dystonia by Transcranial Stimulation," *Annals of Neurology* 75 (2014): 700–7.

53. Ibid.; E. Altenmüller, personal communication, July 1, 2021.

54. Vining, *Notes*, 40.

55. Ibid., 32–40.

56. E. Fishman, "Alex Klein's Improbable Encore," *Chicago Magazine,* January 17, 2017, https://www.chicagomag.com/Chicago-Magazine/February-2017/Oboe-Alex-Klein.

57. J. von Rhein, "How the CSO's Top Oboist Overcame Neurological Disorder to Play Again," *Chicago Tribune,* August 11, 2016, https://www.chicagotribune.com/entert ainment/ct-alex-klein-cso-oboe-20160810-column.html.

58. A. Klein, "Focal Dystonia and Me," *The Double Reed* 39, no. 3 (2016): 48–62.

59. Fishman, "Alex Klein."

60. A. Klein, personal communication, February 21, 2022.

61. N. Skolnik, personal communication, August 21, 2021.

62. E. Altenmüller, personal communication, July 1, 2021.

63. A. Klein, "Focal Dystonia," 62.

Chapter 8

1. Y. Peng, ed., *In Memory of Fei-Ping Hsu: A World Classic Pianist* (Beijing: China Federation of Literary and Art Publishing House, 2002), in Mandarin Chinese, translations made for the author by A. Pusey.

2. C. E. Seashore, *Psychology of Music* (New York: McGraw Hill, 1938), reprinted by Dover Publications, 1967.

3. M. Solomon, "On Beethoven's Creative Process: A Two-Part Invention," *Music and Letters* 61 (1980): 272–83.

4. S. Isserlis, *Robert Schumann's Advice to Young Musicians. Revisited by Steven Isserlis* (London: Faber & Faber, 2016), 31, 66.

5. H. Cowell, "The Process of Musical Creation," *American Journal of Psychology* 37, no. 2 (1926): 233–36.

6. D. Littlejohn, "A West Coaster with a Global Reputation," *Wall Street Journal,* March 2, 2015, https://www.wsj.com/articles/a-west-coaster-with-a-global-reputation-142 5335199.

7. J. Heggie, personal communication, April 9, 2021.

8. W. Gieseking and K. Leimer, *Piano Technique* (New York: Dover Publications, 1972), reprinted from *The Shortest Way to Pianistic Perfection,* by Leimer-Gieseking (Bryn Mawr, PA: Theodore Presser Co., 1932).

9. R. Gerig, *Famous Pianists & Their Technique* (New York, Washington: Robert B. Luce, Inc., 1974), 305.

10. R. Jourdain, *Music, the Brain, and Ecstasy: How Music Captures Our Imagination* (New York: Quill, HarperCollins, 1997), 229.

11. L. Eustis, personal communication, May 26, 2021.

12. "Pioneer Billie Jean King Moved the Baseline for Women's Tennis," *Fresh Air,* host Terry Gross, September 12, 2013, NPR, https://www.npr.org/2013/09/12/221362904/ pioneer-billie-jean-king-moved-the-baseline-for-womens-tennis.

13. T. Orlick and J. Partington, "Mental Links to Excellence," *Sport Psychologist* 2 (1988): 105–30.

14. C. Clarey, "Olympians Use Imagery as Mental Training," *New York Times*, February 22, 2014, https://www.nytimes.com/2014/02/23/sports/olympics/olympians-use-imagery-as-mental-training.html.

15. Ibid.

16. N. Dijkstra, S. E. Bosch, and M. A. J. van Gerven, "Shared Neural Mechanisms of Visual Perception and Imagery," *Trends in Cognitive Sciences* 23, no. 5 (2019): 423–34.

17. J. Pearson, T. Naselaris, E. A. Holmes, and S. M. Kosslyn, "Mental Imagery: Functional Mechanisms and Clinical Applications," *Trends in Cognitive Sciences* 19, no. 10 (2015): 590–602.

18. A. M. Albers, P. Kok, I. Toni, H. C. Dijkerman, and F. P. de Longe, "Shared Representations for Working Memory and Mental Imagery in Early Visual Cortex," *Current Biology* 23, no. 15 (2013): 1427–31.

19. Pearson, "Mental Imagery," 591.

20. Clarey, "Olympians."

21. E. E. Gordon, *Learning Sequences in Music: A Contemporary Music Learning Theory* (Chicago: GIA Publications, 2012), 3–24.

22. B. Green, with W. T. Gallwey, *The Inner Game of Music* (Garden City, NY: Anchor Press/Doubleday, 1986), 60.

23. A. R. Halpern, "Mental Scanning in Auditory Imagery for Tunes," *Journal of Experimental Psychology: Learning, Memory and Cognition* 14 (1988): 434–43; A. R. Halpern, "Perceived and Imagined Tempos of Familiar Songs," *Music Perception* 6 (1988): 193–202; A. R. Halpern, R. Zatorre, M. Bouffard, and J. Johnson, "Behavioral and Neural Correlates of Perceived and Imagined Musical Timbre," *Neuropsychologia* 42 (2004): 1281–92.

24. B. J. Lucas, E. Schubert, and A. R. Halpern, "Perception of Emotion in Sounded and Imagined Music," *Music Perception* 27, no. 5 (2010): 399–412.

25. C. F. Lima, N. Lavan, S. Evans, Z. Agnew, A. R. Halpern, P. Shanmugalingam, S. Meekings, and D. Boebinger, "Feel the Noise: Relating Individual Differences in Auditory Imagery to the Structure and Function of Sensorimotor Systems," *Cerebral Cortex* 25 (2015): 4638–50; R. J. Zatorre and A. R. Halpern, "Mental Concerts: Musical Imagery and Auditory Cortex," *Neuron* 47 (2005): 9–12; S. C. Herholz, A. R. Halpern, and R. J. Zatorre, "Neuronal Correlates of Perception, Imagery, and Memory for Familiar Tunes," *Journal of Cognitive Neuroscience* 24 (2012):1382–97.

26. A. R. Halpern and R. J. Zatorre, "When That Tune Runs through Your Head: a PET Investigation of Auditory Imagery for Familiar Melodies," *Cerebral Cortex* 9 (1999): 697–704.

27. B. Kleber, N. Birbaumer, R. Veit, T. Trevorrow, and M. Lotze, "Overt and Imagined Singing of an Italian Aria," *Neuroimage* 36 (2007): 889–900.

28. The Kennedy Center, "What Goes On inside the Brain of a World Class Singer?," YouTube video, 3:08, January 13, 2018, https://www.youtube.com/watch?v=7nS4u84ppf4.

29. Gordon, *Learning Sequences*.

30. A. Pascual-Leone, N. Dang, L. G. Cohen, J. P. Brasil-Neto, A. Cammarota, and M. Hallett, "Modulation of Muscle Responses Evoked by Transcranial Magnetic

Stimulation during the Acquisition of New Fine Motor Skills," *Journal of Neurophysiology* 74, no. 3 (1995): 1037–45.

31. Z. Highben and C. Palmer, "Effects of Auditory and Motor Mental Practice in Memorized Piano Performance," *Bulletin of the Council for Research in Music Education* 159 (2004): 58–65; I. G. Meister, T. Krings, H. Foltys, B. Boroojerdi, M. Müller, R. Töpper, and A. Thron, "Playing Piano in the Mind—An fMRI Study on Music Imagery and Performance in Pianists," *Cognitive Brain Research* 19 (2004): 219–28; R. Kristeva, V. Chakarov, J. Schulte-Mönting, and J. Spreer, "Activation of Cortical Areas in Music Execution and Imagining: A High-Resolution EEG Study," *NeuroImage* 20 (2003): 1872–83; F. J. P. Langheim, J. H. Callicott, V. S. Mattay, J. H. Duyn, and D. R. Swinberger, "Cortical Systems Associated with Covert Music Rehearsal," *NeuroImage* 16 (2002): 901–8; P. Mikzsa, "The Effect of Mental Practice on the Performance Achievement of High School Trombonists," *Contributions to Music Education* 32, no. 1 (2005): 75–93; D. Cahn, "The Effects of Varying Ratios of Physical and Mental Practice, and Task Difficulty on Performance of a Tonal Pattern," *Psychology of Music* 36, no. 2 (2008): 179–91.

32. B. C. Clark, N. K. Mahato, M. Nakazawa, T. D. Law, and J. S. Thomas, "The Power of the Mind: The Cortex as a Critical Determinant of Muscle Strength/Weakness," *Journal of Neurophysiology* 112 (2014): 3219–26.

Chapter 9

*. H. Pleasants, trans and ed., *Schumann on Music: A Selection from the Writings* (New York: Dover Publications, 2012), 157–58.

1. C.-J. Tsay, "Sight over Sound in the Judgment of Music Performance," *Proceedings of the National Academy of Sciences* 110, no. 36 (2013): 14580–85.

2. C.-J. Tsay, "The Vision Heuristic: Judging Music Ensembles by Sight Alone," *Organizational Behavior and Human Decision Processes* 124, no. 1 (2014): 24–33.

3. E. Hatfield, J. T. Cacioppo, and R. L. Rapson, "Emotional Contagion," *Current Directions in Psychological Science* 2, no. 3 (1993): 96–99; B. Wicker, C. Keysers, J. Plailly, J.-P. Royet, V. Gallese, and G. Rizzolatti, "Both of Us Disgusted in *My* Insula: The Common Neural Basis of Seeing and Feeling Disgust," *Neuron* 40, no. 3 (2003): 655–64.

4. T. Adorno, *Minima Moralia: Reflections from Damaged Life*, trans. E. F. N. Jephcott (London: Verso, 2005), 154.

5. M. Iacoboni, *Mirroring People: The New Science of How We Connect with Others* (New York: Farrar, Straus & Giroux, 2008).

6. A. N. Meltzoff and M. K. Moore, "Imitation of Facial and Manual Gestures by Human Neonates," *Science* 198, no. 4312 (1077): 74–78.

7. G. di Pellegrino, L. Fadiga, L. Fogassi, V. Gallese, and G. Rizzolatti, "Understanding Motor Events: A Neurophysiological Study," *Experimental Brain Research* 91 (1992): 176–80; V. Gallese, L. Fadiga, L. Fogassi, and G. Rizzolatti, "Action Recognition

in the Premotor Cortex," *Brain: A Journal of Neurology* 119 (1996): 593–609; E. Kohler, C. Keysers, M. A. Umiltà, L. Fogassi, V. Gallese, and G. Rizzolatti, "Hearing Sounds, Understanding Actions: Action Representation in Mirror Neurons," *Science* 297, no. 5582 (2002): 846–48.

8. L. Fogassi, P. F. Ferrari, B. Gesierich, S. Rozzi, F. Chersi, and G. Rizzolatti, "Parietal Lobe: From Action Organization to Intention Understanding," *Science* 308, no. 5722 (2005): 662–67.

9. G. Rizzolatti, L. Fogassi, and V. Gallese, "Neurophysiological Mechanisms Underlying the Understanding and Imitation of Action," *Nature Reviews Neuroscience* 2, no. 9 (2001): 661–70; G. Rizzolatti and L. Craighero, "The Mirror-Neuron System," *Annual Review of Neuroscience* 27 (2004): 169–92.

10. L. Fadiga, L. Fogassi, G. Pavesi, and G. Rizzolatti, "Motor Facilitation during Action Observation: A Magnetic Stimulation Study," *Journal of Neurophysiology* 73 (1995): 2608–11.

11. R. Hari, N. Forss, S. Avikainen, E. Kirveskari, S. Salenius, and G. Rizzolatti, "Activation of Human Primary Motor Cortex during Action Observation: A Neuromagnetic Study," *Proceedings of the National Academy of Sciences* 95 (1998): 15061–65; M. Iacoboni, R. P. Woods, M. Brass, H. Bekkering, J. C. Mazziotta, and G. Rizzolatti, "Cortical Mechanisms of Human Imitation," *Science* 286 (1999): 2526–28; L. Aziz-Zadeh, S. M. Wilson, G. Rizzolatti, and M. Iacoboni, "Congruent Embodied Representations for Visually Presented Actions and Linguistic Phrases Describing Actions," *Current Biology* 16 (2006): 1818–23; M. Iacoboni, I. Molnar-Szakacs, V. Gallese, G. Buccino, J. C. Mazziotta, and G. Rizzolatti, "Grasping the Intentions of Others with One's Own Mirror Neurons System," *PLoS Biology* 3, no. 3 (2005): e79, https://doi.org/10.1371/journal.pbio.0030079.

12. L. Carr, M. Iacoboni, M.-C. Dubeau, J. C. Mazziotta, and G. L. Lenzi, "Neural Mechanisms of Empathy in Humans: A Relay from Neural Systems for Imitation to Limbic Areas," *Proceedings of the National Academy of Sciences* 100, no. 9 (2003): 5497–502.

13. V. Gallese, C. Keysers, and G. Rizzolatti, "A Unifying View of the Basis of Social Cognition," *Trends in Cognitive Sciences* 8, no. 9 (2004): 396–403; V. Gallese, "The Roots of Empathy: The Shared Manifold Hypothesis and the Neural Basis of Intersubjectivity," *Psychopathology* 36, no. 4 (2003): 171–80; J. A. C. J. Bastiaansen, M. Thioux, and C. Keysers, "Evidence for Mirror Systems in Emotions," *Philosophical Transactions of the Royal Society B* 364 (2009): 2391–404.

14. Iacoboni, *Mirroring People*, 260–61.

15. G. Rizzolatti and C. Sinigaglia, trans. Frances Anderson, *Mirrors in the Brain: How Our Minds Share Actions and Emotions* (Oxford: Oxford University Press, 2008), vii.

16. R. Mukamel, A. D. Ekstrom, J. Kaplan, M. Iacoboni, and I. Fried, "Single-Neuron Responses in Humans during Execution and Observation of Actions, *Current Biology* 20, no. 8 (2010): 750–56.

17. J. Haueisen and T. R. Knösche, "Involuntary Motor Activity in Pianists Evoked by Music Perception," *Journal of Cognitive Neuroscience* 13, no. 6 (2001): 786–92.

18. M. Lotze, G. Scheler, H.-R. M. Tan, C. Braun, and N. Birbaumer, "The Musician's Brain: Functional Imaging of Amateurs and Professionals during Performance and Imagery," *Neuroimage* 20, no. 3 (2003): 1817–29.

19. B. Haslinger, P. Erhard, E. Altenmüller, U. Schroeder, H. Boecker, and A. O. Ceballos-Baumann, "Transmodal Sensorimotor Networks during Action Observation in Professional Pianists," *Journal of Cognitive Neuroscience* 17, no. 2 (2005): 282–93.

20. E. Altenmüller and S. Furuya, "Apollos Gift and Curse: Making Music as a Model for Adaptive and Maladaptive Plasticity," *Neuroforum* 23, no. 2 (2017): A57–A75.

21. G. Buccino and L. Riggio, "The Role of the Mirror Neuron System in Motor Learning," *Kinesiology* 38, no. 1 (2006): 1–13.

22. C. Bruder and C. Wöllner, "Subvocalization in Singers: Laryngoscopy and Surface EMG Effects When Imagining and Listening to Song and Text," *Psychology of Music* 49, no. 3 (2019): 567–80.

23. I. Stravinsky, *An Autobiography* (London: Calder and Boyars, 1975), 53.

24. M. C. Gridley and R. Hoff, "Do Mirror Neurons Explain Misattribution of Emotions in Music?," *Perceptual and Motor Skills* 102, no. 2 (2006): 600–2; H. Shoda, T. Nakamura, M. R. Draguna, S. Kawase, K. Katahira, and S. Yasuda, "Effects of a Pianist's Body Movements on Listeners' Impressions," presented at the Inaugural International Conference on Music Communication Science, Sydney, Australia, December 5–7, 2007, https://citeseerx.ist.psu.edu/viewdoc/download?doi=10.1.1.497.6236&rep=rep1&type=pdf; K. Ohgushi, "Interaction between Auditory and Visual Information in Conveyance of Players' Intentions," *Acoustical Science and Technology* 27, no. 6 (2006): 336–39; K. Katahira, T. Nakamura, S. Kawase, S. Yasuda, H. Shoda, and M. R. Draguna, "The Role of Body Movement in Co-Performers' Temporal Coordination," presented at the Inaugural International Conference on Music Communication Science, Sydney, Australia, December 5–7, 2007, https://citeseerx.ist.psu.edu/viewdoc/summary?doi=10.1.1.526.5966.

25. I. Molnar-Szakacs and K. Overy, "Music and Mirror Neurons: From Motion to 'E'motion," *Social Cognitive and Affective Neuroscience (SCAN)* 1 (2006): 235–41.

26. K. Overy and I. Molnar-Szakacs, "Being Together in Time: Musical Experience and the Mirror Neuron System," *Music Perception* 26, no. 5 (2009): 489–504.

27. A. Tommasini, "Two Pianists: A Virtuoso and a Philosophizer," *New York Times*, July 23, 2009, https://www.nytimes.com/2009/07/24/arts/music/24keyboard.html.

28. Molnar-Szakacs and Overy, "Music and Mirror Neurons," 236.

Chapter 10

*. A. Einstein, quoted in S. Suzuki, *Nurtured by Love: A New Approach to Education* (New York: Exposition Press, 1969), 90.

1. F. H. Rauscher, G. L. Shaw, and C. N. Ky, "Music and Spatial Task Performance," *Nature* 365 (1993): 611.

2. K. Devlin, *Mathematics: The Science of Patterns* (New York: Henry Holt, 1994).

3. C. M. Johnson and J. E. Memmott, "Examination of Relationships between Participation in School Music Programs of Differing Quality and Standardized Test Results," *Journal of Research in Music Education* 54, no. 4 (2006): 293–307.

4. E. Sanders, "Investigating the Relationship between Musical Training and Mathematical Thinking in Children," *Procedia—Social and Behavioral Sciences* 55 (2012): 1134–43.

5. College Board, 2013. *2013 College-Bound Seniors: Total Group Profile Report*, Analysis by Americans for the Arts, 2014, https://www.americansforthearts.org/sites/defa ult/files/pdf/2014/by_program/research__studies_and_publications/one_pagers/ 16.%20SAT%20Scores%202013%20-%20Arts%20Ed.pdf.

6. K. Elpus, "Is It the Music or Is It Selection Bias?: A Nationwide Analysis of Music and Non-Music Students' SAT Scores," *Journal of Research in Music Education* 61, no. 2 (2013): 175–94.

7. M. J. Bergee and K. M. Weingarten, "Multilevel Models of the Relationship between Music Achievement and Reading and Math Achievement," *Journal of Research in Music Education* 68, no. 4 (2012): 398–418.

8. M. Guhn, S. D. Emerson, and P. Gouzouasis, "A Populations-Level Analysis of Associations between School Music Participation and Academic Achievement," *Journal of Educational Psychology* 112, no. 2 (2020): 308–28.

9. C. Y. Wan and G. Schlaug, "Music Making as a Tool for Promoting Brain Plasticity across the Life Span," *The Neuroscientist* 16, no. 5 (2010): 566–77.

10. L. Hetland, "Learning to Make Music Enhances Spatial Reasoning," *Journal of Aesthetic Education* 23, no. 3/4 (2000):179–238.

11. R. D. Romine, "From Bassoonist to Nobel Laureate: An Interview with Thomas Südhof," *The Double Reed* 36, no. 4 (2013): 54–58.

12. A. Blair and R. P. Razza, "Relating Effortful Control, Executive Function, and False Belief Understanding to Emerging Math and Literacy Ability in Kindergarten," *Child Development* 78, no. 2 (2007): 647–63; F. J. Morrison, C. C. Ponitz, and M. M. McClelland, "Self-Regulation and Academic Achievement in the Transition to School," in *Human Brain Development: Child Development at the Intersection of Emotion and Cognition*, ed. S. D. Calkins and M. A. Bell (Washington, DC: American Psychological Association 2010), 203–24.

13. A. L. Duckworth and M. E. P. Seligman, "Self-Discipline Outdoes IQ in Predicting Academic Performance of Adolescents," *Psychological* Science 16 (2005): 939–44.

14. A. Diamond, "Executive Functions," *Annual Review of Psychology* 64 (2013): 135–68; A. Diamond, "Want to Optimize Executive Functions and Academic Outcomes? Simple, Just Nourish the Human Spirit," *Minnesota Symposium on Child Psychology* 37 (2014): 205–32.

15. J. Zuk, C. Benjamin, A. Kenyon, and N. Gaab, "Behavioral and Neural Correlates of Executive Functioning in Musicians and Non-Musicians," *PLoS One* (2014), https:// doi.org/10.1371/journal.pone.0099868.

16. J. J. Hudziak, M. D. Albaugh, S. Ducharme, S. Karama, M. Spottswood, E. Crehan, A. C. Evans, K. N. Botteron, et al., "Cortical Thickness Maturation and Duration of Music Training: Health-Promoting Activities Shape Brain Development," *Journal of the American Academy of Child & Adolescent Psychiatry* 53, no. 11 (2014): 1153–61.

17. M. Norgaard, L. A. Stambaugh, and H. McCranie, "The Effect of Jazz Improvisation Instruction on Measures of Executive Function in Middle School Band Students," *Journal of Research in Music Education* 67, no. 3 (2019): 339–54.

18. S. Anderson and N. Kraus, "Neural Encoding of Speech and Music: Implications for Hearing Speech in Noise," *Seminars in Hearing* 32 (2011): 129–41.

19. N. Kraus and T. White-Schwoch, "Unraveling the Biology of Auditory Learning: A Cognitive-Sensorimotor-Reward Framework," *Trends in Cognitive Sciences* 19, no. 11 (2015): 642–54.

20. Ibid.

21. P. M. Picciotti, R. Bussu, L. Calò, R. Gallus, E. Scarano, G. Di Cintio, F. Cassarà, and L. D'Alatri, "Correlation between Musical Aptitude and Learning Foreign Languages: An Epidemiological Study in Secondary School Italian Students," *Acta Italica Otorhinolaryngologica* 38, no. 1 (2018): 51–55; K. M. Ludke, F. Ferreira, and K. Overy, "Singing Can Facilitate Foreign Language Learning," *Memory & Cognition* 42 (2014): 41–52.

22. Anderson and Kraus, "Neural Encoding."

23. N. Kraus, E. Skoe, A. Parbery-Clark, and R. Ashley, "Experience-Induced Malleability in Neural Encoding of *Pitch, Timbre,* and *Timing,*" *Annals of the New York Academy of Sciences* 1169 (2009): 543–57.

24. N. Kraus and T. Nicol, "The Musician's Auditory World," *Acoustics Today* 6, no. 3 (2010): 15–27.

25. Kraus and White-Schwoch, "Unraveling the Biology"

26. Anderson and Kraus, "Neural Encoding"

27. A. D. Patel, "Why Would Musical Training Benefit the Neural Encoding of Speech?: The OPERA Hypothesis," *Frontiers in Psychology* 2 (2011), https://doi.org/10.3389/fpsyg.2011.00142.

28. N. Kraus, "Memory for Sound: The BEAMS Hypothesis [Perspective]," *Hearing Research* 407 (2021), https://doi.org/10.1016/j.heares.2021.108291.

29. Kraus and White-Schwoch, "Unraveling the Biology," 29.

30. W. Nager, C. Kohlmetz, E. Altenmüller, A. Rodriguez-Fornells, and T. F. Münte, "The Fate of Sounds in Conductors' Brains: An ERP Study," *Brain Research, Cognitive Brain Research* 17, no. 1 (2003): 83–93.

31. M. Seppänen, E. Brattico, and M. Tervaniemi, "Practice Strategies of Musicians Modulate Neural Processing and the Learning of Sound-Patterns," *Neurobiology of Learning and Memory* 87 (2007): 236–47.

32. G. Musacchia, M. Sams, E. Skoe, and N. Kraus, "Musicians Have Enhanced Subcortical Auditory and Audiovisual Processing of Speech and Music," *Proceedings of the National Academy of Sciences* 104, no. 40 (2007): 15894–98.

33. K. Woodruff Carr, T. White-Schwoch, A. T. Tierney, D. L. Strait, and N. Kraus, "Beat Synchronization Predicts Neural Speech Encoding and Reading Readiness in Preschoolers," *Proceedings of the National Academy of Sciences* 111, no. 40 (2014): 14559–64; N. Kraus and S. Anderson, "Beat-Keeping Ability Relates to Reading Readiness," *The Hearing Journal* (March 2015): 54–55.

34. A. Tierney and N. Kraus, "The Ability to Move to a Beat Is Linked to the Consistency of Neural Responses to Sound," *Journal of Neuroscience* 33, no. 38 (2013): 14981–88.

35. A. Parbery-Clark, E. Skoe, and N. Kraus, "Musical Experience Limits the Degradative Effects of Background Noise on the Neural Processing of Sound," *Journal of Neuroscience* 28, no. 45 (2009): 14100–7.

36. N. Kraus and B. Chandrasekaran, "Music Training for the Development of Auditory Skills," *Nature Reviews Neuroscience* 11 (2010): 599–605.

37. D. L. Strait, N. Kraus, E. Skoe, and R. Ashley, "Musical Experience Promotes Subcortical Efficiency in Processing Emotional Vocal Sounds," *Annals of the New York Academy of* Sciences 1169 (2009): 209–13.

38. M. Geretsegger, C. Elefant, K. A. Mössler, and C. Gold, "Music Therapy for People with Autism Spectrum Disorder," *Cochrane Database of Systematic Reviews* (June 17, 2014), https://doi.org/10.1002/14651858.CD004381.pub3; E.-M. Quintin, "Music-Evoked Reward and Emotion: Relative Strengths and Response to Intervention of People with ASD," *Frontiers in Neural Circuits* 13, no. 49 (2019): https://doi.org/10.3389/fncir.2019.00049.

39. A. Parbery-Clark, D. L. Strait, S. Anderson, E. Hittner, and N. Kraus, "Musical Experience and the Aging Auditory System: Implications for Cognitive Abilities and Hearing Speech in Noise," *PLoS One* 6, no. 5 (2011), https://doi.org/10.1371/journal.pone.0018082.

40. T. White-Schwoch, K. Woodruff Carr, S. Anderson, D. L. Strait, and N. Kraus, "Older Adults Benefit from Music Training Early in Life: Biological Evidence for Long-Term Training-Driven Plasticity," *Journal of Neuroscience* 33, no. 45 (2013): 17667–74.

41. M. E. Ellwood-Lowe, R. Foushee, and M. Srinavasan, "What Causes the Word Gap?: Financial Concerns May Systematically Suppress Child-Directed Speech," *Developmental Science* (July 8, 2021), https://doi.org/10.1111/desc.13151.

42. A. Loughan and R. Perna, "Neurocognitive Impacts for Children of Poverty and Neglect," *American Psychological Association CYF News* (July 2012), https://www.apa.org/pi/families/resources/newsletter/2012/07/neurocognitive-impacts.

43. A. E. Margolis, B. Ramphal, D. Pagliaccio, S. Banker, E. Selmanovic, L. V. Thomas, P. Factor-Litvak, F. Perera, et al., "Prenatal Exposure to Air Pollution Is Associated with Childhood Inhibitory Control and Adolescent Academic Achievement," *Environmental Research* 202, no. 7 (2021): https://doi.org/10.1016/j.envres.2021.111570.

44. E. Skoe, J. Krizman, and N. Kraus, "The Impoverished Brain: Disparities in Maternal Education Affect the Neural Response to Sound," *Journal of Neuroscience* 33, no. 44 (2013): 17221–31.

45. "New Report Finds Many Families with Children Are Depressed, Uninsured, Hungry and at Risk of Foreclosure or Eviction," posted December 14, 2020, https://www.aecf.org/blog/new-report-finds-many-families-with-children-are-depressed-uninsured-hungry.

46. N. Kraus, personal communication, July 29, 2019.

47. A. T. Tierney, J. Krizman, and N. Kraus, "Music Training Alters the Course of Adolescent Auditory Development," *Proceedings of the National Academy of Sciences* 112, no. 32 (2015): 10062–67.

48. M. Martin, personal communication, February 2, 2021.

49. N. Kraus, J. Slater, E. C. Thompson, J. Hornickel, D. L. Strait, T. Nicol, and T. White-Schwoch, "Music Enrichment Programs Improve the Neural Encoding of Speech in At-Risk Children," *Journal of Neuroscience* 34, no. 36 (2014): 11913–18.

50. M. Martin, personal communication, February 2, 2021.

51. H. M. Holbrook, M. Martin, D. Glik, J. J. Hudziak, W. E. Copeland, C. Lund, and J. D. Fender, "Music-Based Mentoring and Academic Improvement in High-Poverty Elementary Schools," *Journal of Youth Development* 17 (March 2022), https://jyd.pitt.edu/ojs/jyd/article/view/221701FA2.

52. N. Rabkin and E. C. Hedberg, "Arts Education in America: What the Declines Mean for Arts Participation, Based on the 2008 Survey of Public Participation in the Arts," Research Report #52 (Washington, DC: National Endowment for the Arts, 2011).

Epilogue

1. "'This Is Why We Play': Amid Pandemic, Philadelphia Orchestra Livestreams Beethoven," *Fresh Air*, with Terry Gross, National Public Radio, Philadelphia: WHYY, March 19, 2020, https://www.npr.org/2020/03/19/817760744/this-is-why-we-play-amid-pandemic-philadelphia-orchestra-livestreams-beethoven.

2. A. Anshel and D. A. Kipper, "The Influence of Group Singing on Trust and Cooperation," *Journal of Music Therapy* 25, no. 3 (1988): 145–55.

3. C. Grape, M. Sandgren, L.-O. Hansson, M. Ericson, and T. Theorell, "An Empirical Study of Professional and Amateur Singers during a Singing Lesson," *Integrative Physiological and Behavioral Science* 38, no. 1 (2003): 65–75; J. R. Keeler, E. A. Roth, B. L. Neuser, J. M. Spitsbergen, D. J. M. Waters, and J.-A. Vianney, "The Neurochemistry and Social Flow of Singing: Bonding and Oxytocin," *Frontiers in Human Neuroscience* 9 (2015), https://doi.org/10.3389/fnhum.2015.00518.

4. G. Kreutz, "Does Singing Facilitate Social Bonding?," *Music & Medicine* 6, no. 2 (2014): 51–60.

5. V. N. Salimpoor, M. Benovoy, K. Larcher, A. Dagher, and R. J. Zatorre, "Anatomically Distinct Dopamine Release during Anticipation and Experience of Peak Emotion to Music," *Nature Neuroscience* 14, no. 2 (2011): 257–62; D. M. Greenberg, J. Decety, and I. Gordon, "The Social Neuroscience of Music: Understanding the Social Brain through Human Song," *American Psychologist* (2021), https://doi.org/10.1037/amp0000819.

6. D. Fancourt and A. Williamon, "Attending a Concert Reduces Glucocorticoids, Progesterone and the Cortisol/DHEA Ratio," *Public Health* 132 (2016): 101–4.

7. U. Lindenberger, S.-C. Li, W. Gruber, and V. Müller, "Brains Swinging in Concert: Cortical Phase Synchronization While Playing Guitar," *BMC Neuroscience* 10, no. 22 (2009), https://doi.org/10.1186/1471-2202-10-22.

8. J. Grahn and M. Henry, "What Makes Musical Rhythm Special: Cross-Species, Developmental, and Social Perspectives," Presented at a meeting of the Cognitive Neuroscience Society, Boston, March 27, 2018.

9. Y. Hou, B. Song, Y. Hu, Y. Pan, and Y. Hu, "The Averaged Inter-Brain Coherence between the Audience and a Violinist Predicts the Popularity of Violin Performance," *NeuroImage* 211 (2020), https://doi.org/10.1016/j.neuroimage.2020.116655.

10. I. Kokal, A. Engel, S. Kirschner, and C. Keysers, "Synchronized Drumming Enhances Activity in the Caudate and Facilitates Prosocial Commitment—If the Rhythm Comes Easily," *PLoS One* 6, no. 11 (2011), https://doi.org/10.1371/journal.pone.0027272.

11. S. S. Wiltermuth and C. Heath, "Synchrony and Cooperation," *Psychological Science* 20, no. 1 (2009), https://doi.org/10.1111/j.1467-9280.2008.02253.x.

12. S. Kirschner and M. Tomasello, "Joint Music Making Promotes Prosocial Behavior in 4-Year-Old Children," *Evolution and Human Behavior* 31 (2010): 354–64.

13. L. K. Cirelli, K. M. Einarson, and L. J. Trainor, "Interpersonal Synchrony Increases Prosocial Behavior in Infants," *Developmental Science* 17, no. 6 (2014): 1003–11.

14. D. Jacobsen, "Leonard Hussey's Banjo: Brain Food," accessed May 25, 2021, https://www.oceanwide-expeditions.com/blog/leonard-hussey-s-banjo-brain-food.

15. J. Palant, personal communication, August 23, 2021.

16. A. Nordberg, C. Cronley, E. Murphy, C. Keaton, and J. Palant, "The Dallas Street Choir: The Impact of Communal Singing on Those Experiencing Homelessness," *Choral Journal* 59, no. 3 (2018): 8–20.

17. Ibid., 13.

18. M. Martin, personal communication, September 9, 2021.

19. J. Palant, personal communication, Aug 23, 2021.

20. D. Barenboim, "West-Eastern Divan Orchestra," accessed August 20, 2021, https://west-eastern-divan.org/founders/daniel-barenboim.

Glossary

3-D motion capture A process of recording the movement of objects and people that is used in video game design, animation, robotics, and movement sciences. In music research, it can be used to map movement time onto musical time, showing how a musician moves at his instrument, for example.

Absolute pitch The ability to identify a pitch or to produce a pitch designated by name without first hearing a reference tone, commonly called "perfect pitch."

Alexander technique A century-old mind-body practice developed by Frederick Matthias Alexander. The technique teaches awareness of the functioning of one's nervous, muscular, and skeletal systems to alleviate habitual patterns of tension, allowing one to rediscover the body's natural balance.

Amusia Sometimes referred to as "tone deafness," amusia is the inability to produce or perceive pitch accurately and can either be congenital or the result of brain damage. Affecting about 4 percent of the population, it can also extend to impairments in processing timbre, emotion, musical memory, and the ability to tap in time to music, known as beat deafness. Amusics are still able to speak and understand speech.

Aphantasia The inability to form mental images in one's mind, estimated to affect 1–4 percent of the population.

Audiation A term coined by music educator Edwin Gordon to refer to the ability to hear and comprehend music internally when no sound is physically present. It is a cognitive process by which the brain gives meaning to musical sounds, and Gordon suggested that audiation is to music what thought is to language.

Beat deafness The inability to distinguish musical rhythm or sync to a beat. It is considered a form of congenital amusia.

Beat induction The ability to detect the beat in music. It is already functional right after birth.

Consonance and dissonance Subjective qualities that are assigned to music intervals. Generally, consonant intervals have been considered as sounding "pleasant," and stable, with no sense of motion; dissonant intervals as "unpleasant" or unstable and needing to resolve. But what is considered dissonant in one culture may be considered consonant in another. Even within the Western tradition, parallel ninth or seventh chords, which would at one time have been considered dissonant, are now seen as a color or texture device in Debussy's music.

Contextual interference effect An effect that occurs in learning when one practices different skills or different kinds of music at the same time and alternates between them.

The interference that results slows down learning—resulting in poorer performance in the short term, but in superior retention over the long term.

Desirable difficulty A learning strategy that requires considerable effort, slowing down the learning process, but leading to the opportunity to add more relevant information and strengthening long-term retention. Introduced by Robert Bjork and Elizabeth Ligon Bjork in the early 1990s.

Electroencephalography, EEG A non-invasive technique that can detect and record changes in electrical activity in the brain through electrodes placed on the scalp. It produces a chart that shows brain wave activity. It is used to diagnose brain disorders such as epilepsy and is also used in cognitive development research.

Embodied cognition Knowing with the body. As opposed to traditional theories that see cognition as confined to the brain, embodied cognition sees motor and sensory systems as integrated with cognitive processing.

Embodied music cognition Perceiving and understanding music through the body; giving meaning to music through movement.

Embodied simulation The possibility that we understand movement behavior of someone we are observing by simulation in our own body. It has been proposed that mirror neurons provide the mechanism for embodied simulation, for us to understand the behavior or actions of others, by mapping that behavior or actions onto our own motor system.

Emotional contagion Spontaneous spread of emotions or behaviors from one person to another or within a group—helps synchronize our emotions with others and is linked to empathy. Getting caught up in the emotions of those around you at a concert is emotional contagion, as is becoming depressed if you are surrounded by others who are depressed.

Entrain The ability to synchronize to a beat, clap together, or dance to music. Entrainment depends on beat induction, the ability to detect the beat in music.

Frequency In music, the number of cycles per second of the sound wave, measured in Hertz (Hz) and perceived as pitch. The more cycles per second, the higher the pitch. The A just below middle C on the keyboard is A = 220 Hz, the A above middle C is 440 Hz. Brain waves are also measured by number of cycles per second, from the slowest delta waves, at 1–3 Hz, indicating sleep, to the fastest gamma waves, above 30 Hz, indicating high-level information processing.

Functional magnetic resonance imaging, fMRI Measures brain activity by detecting changes in blood oxygenation and flow in response to neural activity. It is used in neuroscience research to see what parts of the brain are functioning when a subject is performing different tasks such as listening to music, singing, or playing an instrument.

Harmonics The frequencies other than the fundamental that are part of any pitch generated on an instrument or with the voice. They are not heard because they are much softer, but they contribute to the richness of the sound. The A below middle C on the keyboard has a frequency of 220 Hz, but it is also vibrating at 440 Hz, 660 Hz, 880 Hz, and higher. Those frequencies are harmonics. Harmonic frequencies have a mathematical relationship to the fundamental. An octave is in a relationship of 2:1 (interval between second and first harmonics), a fifth is 3:2 (the interval between third and second harmonics), and a fourth is 4:3 (the interval between the fourth and third harmonics).

Homunculus Refers to the neural representation of all our body parts in both the sensory cortex and the motor cortex, each part occupying an area of the sensory or motor cortex relative to where we need the most sensitivity (sensory cortex) or the amount of use (motor cortex). Fingers and hands occupy more area in both the sensory and motor cortices than do feet or toes because we use them much more extensively. The neural representation was named "homunculus," a seventeenth-century diminutive for "man," by Wilder Penfield, who discovered the representations in the 1930s and 1940s.

Infant-directed speech, or IDS Sometimes called "baby talk," "motherese," or "parentese." A more musical form of speech used with babies, with exaggerated highs and lows, a slower rate of speaking, long vowels, a larger dynamic range, and made-up words resembling the actual word, such as "da-da" for daddy, or "wa-wa" for water. It is preferred by babies to adult speech.

Kinesthesia The awareness of the position and movement of parts of the body. Kinesthesia is the sense that delivers information to the brain about effort, movement, position, and weight delivery, and is used in motor imagery.

Magnetoencephalography, MEG A neuroimaging technique that measures the magnetic fields produced by the brain's electrical currents. It can be used to locate the source of epileptic seizures and is also used in research settings to map motor and sensory areas, language, vision, and other functions.

Microtones An interval in music smaller than the semitone or half-step found in the Western tuning system of twelve equal intervals per octave. Music of other cultures, such as Indonesian, Middle Eastern, and South Asian, regularly uses microtones of varying sizes.

Musicality Term adopted by origin-of-music researchers to refer to the evolutionary brain processes that support musical behavior, such as being able to pick up a beat or recognize a tune no matter what pitch it begins on. For musicians, musicality is the ability to communicate the emotional essence of a work and to bring expressiveness to performance.

Neural crosstalk Also called interlimb skill transfer. Learning a motor skill in one limb can transfer to the opposite limb via the corpus callosum; one hemisphere of the brain is teaching the other hemisphere.

Notational audiation The ability to silently read a musical score, hear it in one's mind, and give musical meaning to the notation.

Novelty preference In general, very young infants prefer the familiar, whether sounds or faces. But as the infant becomes older, after seeing or hearing the familiar, he will turn to something new.

Positron emission tomography, PET A type of imaging scan that uses small amounts of radioactive materials called "tracers" to evaluate organ and tissue functions. PET is commonly used to detect cancer and to evaluate cancer treatment. It is also a research tool to learn more about the functioning of the human brain.

Proprioception Your awareness of your body's position in space. Often used interchangeably with kinesthesia, but they are not the same. Proprioception is more cognitive, focusing on awareness of body's position; kinesthesia is more behavioral, focusing on movement or weight delivery.

Proto-musical language Researchers have proposed that this form of communication varying in pitch, timing, and timbre was probably used by our prehistoric ancestors to bond—mothers with infants, and members of a group with one another. It was used as a means of communication about daily tasks of survival or withstanding assaults from rival groups. It eventually split into language to express facts and ideas and music to express emotion.

Relative pitch The ability, given a reference pitch, to identify or re-create another pitch and identify the interval between them.

Sound before sign Also called "sound before symbol." The theory that one should be able to hear music in one's mind, to "think in sound," before learning music notation. It is equivalent to understanding and speaking a language before learning to read.

Taubman technique A technique developed by pianist Dorothy Taubman whose aim was to solve physiological problems of piano interpretation through an understanding of the underlying principles of biomechanics and anatomy. The technique is used to treat repetitive strain injuries and other injuries of pianists, as well as to foster a healthy approach to the instrument to avoid injury.

Transcranial direct current stimulation, tDCS A non-invasive, painless treatment that uses direct electrical currents to stimulate specific parts of the brain, used to treat a variety of conditions, including Parkinson's disease, traumatic brain injury, and stroke. Has also been used to treat movement disorders, as in focal dystonia.

Transcranial magnetic stimulation, TMS A noninvasive procedure that uses magnetic fields to stimulate nerve cells in the brain. It has been used as a treatment for depression and anxiety but has also been used in research to learn more about how the brain controls behavior and how the brain is organized.

Index

For the benefit of digital users, indexed terms that span two pages (e.g., 52–53) may, on occasion, appear on only one of those pages.

Note: Figures are indicated by f following the page number